SHEEP MAY SAFELY GRAZE

Schafe können sicher weiden
wo ein guter Hirte wacht,
wo Regenten wohl regieren
kann man (Ruh' und Frieden spüren)
und was Länder glücklich macht.

Flocks and herds may safely pasture
When their shepherds guard them well!
They whose monarch loves them truly,
Knows their needs and fills them duly,
Know and will in peace and concord dwell.

From Henry S. Drinker, *Texts of the Choral Works of Johann Sebastian Bach in English Translation,* printed privately and distributed by the Association of American Colleges Arts Program, 19 West 44th Street, New York City n.d., p. 499.

A Personal Essay on Tradition and a Contemporary Sheep Ranch

Sheep May Safely Graze

BY LOUIE W. ATTEBERY

A Publication in Northwest Folklife,
Louie W. Attebery, General Editor

University of Idaho Press
Moscow, Idaho
1992

Copyright © Louie W. Attebery, 1992
Published by the University of Idaho Press, Moscow, Idaho 83843
No part of this book may be reproduced, stored in a retrieval system, or transmitted in any form or by any means, electronic, mechanical, photocopying, recording, or otherwise, except for use in reviews, without the prior permission of the publisher.

Design by Karla Fromm

Printed in the United States of America
97 96 95 94 93 92   5 4 3 2 1

Material from Robert Frost's "Two Tramps in Mud Time," is taken from *The Poetry of Robert Frost,* edited by Edward Connery Lathem, copyright 1936 by Robert Frost, copyright © 1964 by Lesley Frost Ballantine, copyright © 1969 by Holt, Rinehart and Winston, Inc., and reprinted by permission of Henry Holt and Company, Inc.; this material is also used by permission of the Estate of Robert Frost, Edward Connery Lathem, and Jonathan Cape, Publisher, Random Century Group, London. Material from Alexander Campbell McGregor's *Counting Sheep: From Open Range to Agribusiness in the Columbia Plateau* (Seattle, 1982) is used by permission of the University of Washington Press. Material from Orren McMullen's *Vanishing Kingdom* (Weiser, Idaho, 1972) is used by permission of J.W. McMullen. "Shear Trivia About Sheep," by David Procter, Business Section, 20 September 1987, used by permission of the *Idaho Statesman.* Material from Helen Hooven Santmyer's, *And the Ladies of the Club* (Columbus, Ohio, Berkeley Book, 1982) is used by permission of The Ohio State University Press.

Library of Congress Cataloging-in-Publication Data

Attebery, Louie W. (Louie Wayne), 1927–
  Sheep may safely graze : a personal essay on tradition and a contemporary sheep ranch / by Louie W. Attebery.
    p. cm. — (A Publication in northwest folklife)
  Includes bibliographical references (p.   ) and index.
  ISBN 0-89301-158-4
  1. Sheep ranches—Idaho. 2. Soulen family. 3. Sheep—Idaho.
4. Ranch life—Idaho.   I. Title. II. Series.
SF375.4.I2A88 1992
636.3'009796—dc20   92-26471
                                                                CIP

Contents

ACKNOWLEDGMENTS vii
PREFACE ix
INTRODUCTION BY TERESA SOULEN LITTLE xiii
CHAPTER ONE 1
which discusses transhumance and tradition and establishes the pattern of theme and variation . . .

CHAPTER TWO 25
which treats of shearing, lambing, marking, and associated activities . . .

CHAPTER THREE 59
which focuses upon the life of a sheepherder, how he does his work, and the provisioning processes that sustain him . . .

CHAPTER FOUR 93
which touches upon the fabled hostility between sheep- and cattlemen, meanders over the problem of urban-agrarian perceptions, and concludes with some observations about reality and whether it changes . . .

APPENDIX A 103
APPENDIX B 105
APPENDIX C 111
APPENDIX D 115
APPENDIX E 117
BIBLIOGRAPHY 127
INDEX 129

Acknowledgments

At least to these people and organizations, I owe sincere thanks and appreciation, and the reader is to understand that they bear none of the blame for whatever deficiencies the present study possesses: the Regional Studies Center of Albertson College of Idaho and the secretarial help of Linda Batie; the great kindness and indulgence of Phil Soulen and his family, who had to endure a great many questions and a presence whose greatest concern was often how to avoid getting in the way; and finally those shadowy pioneer sheepmen whose bands passed by a small rural school to the delight and wonderment of the pupils... Tommy Carr, Ivan and George Stover, Barney Linkous, and the Leightons, and the Hands, and the Blevinses....

Preface

Idaho Folklife Associates was the name of a nonprofit corporation composed of persons interested professionally in advancing the cause of folklife studies. At one of its early regular meetings, the problem of appropriate topics for study was discussed, and one of us suggested that certain aspects of our cultural heritage were in perilous condition, either diminishing in importance and declining after the fashion of what cultural evolution chooses to weed out or lingering only in memory or in faint hints on the cultural landscape. Another Associate remarked that concentration upon such vulnerable aspects of regional culture might be confused with simple antiquarianism. In response, yet another Associate defended antiquarianism by saying that to a discerning eye nothing is simple and, moreover, that Alan Jabbour had argued somewhere that all of us interested in folklife are necessarily antiquarian, at least in part. Thus, the antiquarian impulse had been defended by a spokesman of considerable merit.

Finally, a list of regional topics developed, any one of which might be fit and proper for an organization of scholars to study. On the list were these: (1) planned Idaho communities such as New Plymouth, Aurora, and Mormon platted towns like Chesterfield; (2) health and other early resorts and spas, like Starkey, Hot Lake Sanitarium (Oregon), Skippen's Hot Springs, and Campbell's Hot Springs; (3) visible and invisible ethnic and linguistic communities—Finns, Irish, Czechs, Scots, Chinese, blacks, Japanese, Slavs, Dutch, and others; (4) the people who live alone—trappers, prospectors, squatters, hermits; (5) subsistence homesteads; (6) the old "division points" on the Union Pacific Railroad; (7) academies and colleges founded, perhaps, on the Oberlin model—the Intermountain Institute, Gooding College; (8) transhumance.

As we discussed each, the logical and easily the most exciting topic was transhumance, the movement of men and livestock across the landscape in cycles determined by the availability of pasture. We agreed that we should "choose one representative, long-lived, still operating sheep ranch within the study area [Southwest Idaho–Eastern Oregon]" on which to concentrate. Then came the task of identifying such a ranch. Name after name was advanced, and to our surprise and dismay, name following name was abandoned, for "_____ has just sold out," "has just retired," "has converted from sheep to cattle." In this fashion, we learned what I suppose any county agent or a brief survey of the USDA *Agricultural Statistics* could have told us—that sheep raising is declining. More particularly, we would have learned that tending sheep in the traditional pattern of seasonal migration was going downhill fast. We were fortunate to find Phil Soulen, a sheepman whose operation seemed stable and who felt inclined to provide the pattern for the "long-lived, still operating sheep ranch within the study area."

We had not foreseen the paucity of examples. Neither had we foreseen the evolution—if that's what it is—of the research, fieldwork, and writing from a group-centered activity to one for which an individual was responsible. But such a shift may have been for the best, not because of the gifts of the researcher, but simply because he knows the sheepman, knew his father, knows something of the region and its cultural legacy, and had the advantage of growing up on a small ranch that made extensive use of the public domain prior to 1934. My grandfather (it's time to shift from third to first person) even had a small band of sheep when I was a boy.

Several pieces of information give this study more substance through implication than might be readily apparent. Implicit in it, for instance, is a particularized validation of the statistics behind the general phenomenon of rural flight, of fewer and fewer people owning more and more of the land. Sheepmen are going out of business.* In the face of the national statistic that in 1870 50 percent

---

* It is generally agreed that the main reasons for this decline include a reduction in the amount of range necessary to support huge numbers of sheep, the difficulty in finding hired hands willing to live the isolated life of a herder, competition from Australia and New Zealand, and the appearance of inexpensive synthetic textiles.

of our labor force was involved in agriculture and animal husbandry and that exactly 100 years later the figure was 5 percent, the Soulen sheep outfit survives. An item from the *Idaho Statesman* of 10 February 1987 notes that in 1867, when the USDA began keeping records, 45,000,000 head of sheep and lambs were tallied; in 1986 the tally was 10,500,000. An article in the *North American Review* (vol. 147, no. 381:220) of August 1888 estimated the number of sheep at 50,000,000, with the value of the annual product—mutton, wool, increase in flocks—in excess of $125,000,000. And that in 1888 dollars! (The same article says that of the six regions of the United States and its territories, the western states commanded 37 percent of the resource.) All of these somewhat dismal statistics might be compared with those coming from New Zealand (about the size of Colorado), with about 3,000,000 people and 70,000,000 sheep, according to information given on 5 February 1987 on NBC's *Today* show.

As I approached the task of studying a sheep outfit in which context and tradition are vital, it seemed appropriate to make the study personal and informal rather than impersonal, formal, and academic. Thus, the personal pronoun appears, though I hope not oppressively, and academic jargon is minimal. In reading secondary sources about the growth and decline of rangeland sheep operations, I found few that concentrated upon their folklife aspects. Without question, the finest study of the entire matter is a memorable work by a Scots American (that is, an American of Scots heritage) entitled *Counting Sheep: From Open Range to Agribusiness on the Columbia Plateau* (Seattle: University of Washington Press, 1982). Its author is Alexander Campbell McGregor, the product of one of the West's fine liberal arts colleges and of the graduate school of a distinguished state university. Since this is not the place to identify all the sources to which the present study is indebted, I shall mention only *English Creek* by Ivan Doig (New York: Penguin, 1984); *Vanishing Kingdom* by Orren C. McMullen (Weiser, Idaho, 1972), who had the courage and good sense to publish his recollections of early times in Eastern Oregon and Southwestern Idaho; and three works by Harold Lenoir Davis: *Honey in the Horn* (New York: William Morrow, 1935), *Winds of Morning* (New York: William Morrow, 1952), and *Team Bells Woke Me, and Other*

Stories (New York: William Morrow, 1953). But most of the substance of the present work is the result of interviews and direct observations. There is—and I dislike the term almost as much as I dislike the word "disinformation"—some "hands on" experience and insight emerging from sleeping in a herder's tent, eating sheep-camp food, counting sheep, and loading them into trucks. Holdings in the Albertson College of Idaho Folklore Archives were also utilized as were the Idaho State Historical Society's oral history resources, especially the Stewart Cruickshank interviews. A selection appears in Appendix E.

The study is enhanced by the Introduction, prepared by a granddaughter of the founder of the Soulen Livestock Company. Teresa Soulen Little has prepared not only a brief biography of her grandfather but also a careful record of the evolution of the company, truly an institution that is the lengthened shadow of the man, as Emerson might have put it. The first chapter outlines the scope and intent of the study and asks the reader to see it as a blend of tradition and context. The various chapters consist of description and observation, of assessment and interpretation, of transcripts of interviews (when appropriate), of photographs, and of personal insights and comments stimulated by reflections on the passing of time and the need for tradition-inspired stability.

INTRODUCTION
BY TERESA SOULEN LITTLE

Harry Boone Soulen, the oldest child of Philip Hendrick Soulen and Hendrika Boone Soulen, was born on 19 June 1893, probably at Oregon City, Oregon, although a letter to him dated 1970 states that there was no birth record on file for Harry B. Soulen of Oregon City, Oregon. It is possible that he was born at the home of his maternal grandparents, from whom he received his middle name, in Holland, Michigan. His sister Beth was born there.

His early childhood, 1894–1906, was spent in Orange City, Iowa. From there the family moved to Moscow, Idaho, in 1906. The author Carol Ryrie Brink was a playmate of the Soulen children and of the youngsters of another neighborhood family named David. The main character in her book *Louly* is modeled after Beth Soulen; Ko-Ko is Harry.

After attending the Preparatory School of the University of Idaho, graduating on 11 June 1910, he matriculated at the University of Idaho, where he was a member of such organizations as the Agriculture Club, College Orchestra, Biology Club, Phi Delta Theta fraternity (the local fraternity was begun 31 December 1908), and Tau Alpha (Junior Honor Society), winning "A" honors his junior year. He played varsity basketball for three years, lettering as a junior and being elected captain for his senior year by a unanimous vote. Somehow he found time to play varsity tennis his junior year (his doubles partner—and frequent singles opponent—was Don David); to help organize the Iota Alpha fraternity, an honorary fraternity for agriculture; to serve as a member of the livestock judging team for the University of Idaho that won at the Pacific International Livestock show; and to earn a place on the Panhellenic Council his senior year.

From 4 September 1911, until 28 October 1911, as a private of

Company E, Second Infantry Regiment of the National Guard, Harry attended Fort George Wright in Washington.

Following graduation from the University of Idaho on 10 June 1914, with a Bachelor of Science degree in agriculture, specializing in animal husbandry, Harry secured a teaching position at the Beaverhead County High School in Dillon, Montana, where he taught for two years. Here, for a salary of $1,200, he taught the ag classes and coached a winning basketball team. While at Dillon, he was put in charge of all aspects of public relations work for a sheep and wool railroad exhibition car developed by the U.S. Department of Agriculture and the state extension service. It was here that he became a Mason. In 1917 he took out some land under the Homestead Act near Dillon, a claim he relinquished 1 June 1922.

Harry helped organize and was a member of the Spokane Military Band (he had played clarinet earlier for the Moscow City Band), which was sent intact to the United States Navy in August 1917. His navy career was as follows:

Goat Island, San Francisco	12 Aug. 1917–3 Sept. 1917
USS *St. Louis*	3 Sept. 1917–14 Aug. 1918
Peham Bay	14 Aug. 1918–9 Sept. 1918
Off. Com. 3rd Naval Dist.	9 Sept. 1918–21 Oct. 1918
Annapolis	21 Oct. 1918–31 Jan. 1919 (date training was completed)
16th Naval Dist., Seattle	31 Jan. 1919–15 Feb. 1919

It was a career that saw him rise from enlisted man to officer by virtue of an officers' training program at Annapolis, in the first month of which he stood 46th in a class of 450. The USS *St. Louis* was in the Atlantic fleet when Harry was stationed aboard her, and his letters home provide a fascinating picture of life on a navy vessel during what is still referred to as the Great War.

After his return from the navy, Harry went to work for a sheep company in Montana until it went broke. In 1923 he became the county agricultural agent for Washington County in Idaho, filling a position which was terminated later in the year in spite of a 400-signature petition to the Board of County Commissioners asking for appropriations to continue the post.

INTRODUCTION   XV

Harry then went to work for Swift & Company in Colorado, returning to Idaho, where on 31 July 1928, on one of the truly significant occasions of his life, he married Beulah May Johnson in McCall, Idaho. Then he and his brother-in-law, Enderse Van Hoesen, became partners in a sheep outfit. As the operation expanded, capital was supplied by the Boise City Bank, which by September of 1929 was holding a mortgage on their operation for $20,700.

On 5 September 1929, the Mesa Sheep Company formed as a private corporation, acquiring the Jim (James E.) Clinton Crane Creek sheep outfit, which had gone into receivership to the Portland Cattle Loan Company, Inc. The five directors of the company were as follows: H. B. Soulen, president of the corporation; H. B. Duff; E. G. Van Hoesen, vice-president; H. A. Ebling, secretary-treasurer; and John Stringer. The company formed with 500 shares at a value of $100 each, with stock distribution as follows: H. B. Soulen, 83 1/2 shares; E. G. Van Hoesen, 83 1/2 shares; H. A. Ebling, 1 share; John Stringer, 1 share; H. B. Duff, 1 share; H. B. Duff & Company, 330 shares.

On 23 September 1929, the Mesa Sheep Company became mortgaged to H. B. Duff & Company for $115,000 on the following:

Crane Creek Inventory

7,719 ewes 2's	@ 14.00	$108,066.00
51 wethers	@ 7.00	357.00
57 aged ewes	@ 1.00	57.00
113 bucks	@ 30.00	3,390.00
8 bucks	@ 15.00	120.00
		$111,990.00

Turned in by Soulen

2,183	ewes 2's & 3's
25	bucks
454	aged ewes
10,610	

All of the above were branded XX or ⊢⊢ (4 plus connected). Included were 3,000 tons of hay and equipment. The mortgage, which covered all sheep and equipment owned by Mesa Sheep Company, was payable 23 March 1930 at 7 percent interest. The following were also in the Crane Creek inventory: thirteen mules, twelve horses, twenty dogs, one burro, nine camp outfits, sixteen packsaddles, twelve pack bags, four riding saddles, four beds, valued at $2,500.

Later, Duff was to release Mesa Sheep Company from indebtedness for $28,100, to be paid by future earnings of the company and the 1932 crop of wool.

In 1933 the Mesa Sheep Company and Western Properties, Inc. (the financial arm of Union Stock Yards of Chicago), entered a $45,000 mortgage on the land, with Western Properties holding the following stock as collateral: certificate number 8, Mrs. E. G. Van Hoesen, 1 share; number 9, Mr. E. G. Van Hoesen, 83 1/2 shares; number 10, Mr. E. G. Van Hoesen, 165 1/2 shares; number 11, Mr. H. B. Soulen, 166 1/2 shares; number 12, Mr. H. B. Soulen, 82 1/2 shares; number 13, Mrs. H. B. Soulen, 1 share, for a total of 500 shares.

On 17 April 1934, an agreement was made that when the note for the stock was paid, the following certificates were to be delivered to John Stringer: number 9, E. G. Van Hoesen, 83 1/2 shares; number 11, H. B. Soulen, 166 1/2 shares.

On 24 March 1936, an agreement with John Stringer was written that, with the payment of debts (without borrowing or liquidating assets) owed by Mesa to the Idaho Livestock Product Credit Association, stock would be reissued with John Stringer receiving 250 shares.

On 8 January 1937, Harry's brother-in-law Don David sent a check for $25,000 from Beth's trust fund to Harry, who then invested it in stock in her name, Elizabeth Soulen David. Don suggested that a corporation should continue. On 12 January 1937 Harry agreed to pay Stringer $25,000 with 5 percent interest, an arrangement making Harry owner of 50 percent of the stock. The stock note was to be paid on that 50 percent with shares of stock from Mesa Sheep Company. On 11 December 1939 an agreement was made with Stringer and Duff, through which Stringer held a note from Harry with Duff, the owner of one-half interest in it. Eleven months later, on 11 November 1940, Harry settled with Duff and Stringer for $22,500, saving him $2,500 on the note and $3,525 in interest. By 19 December 1940, Beth owned one-quarter interest, Van one-quarter interest, and Harry one-half interest in Mesa Sheep Company, a corporation.

In 1941 it was changed to a partnership at Don David's suggestion to avoid the terrific tax penalties to which the corporation was subjected. The firm was reincorporated as Soulen Livestock Company in early 1965.

Harry served as vice-president of the Idaho Wool Growers for five years before being elected president on 5 January 1940, a position he held through 1943. The following appreciation to him is from Dockey* Bettis, dated 17 January 1942:

> This is to thank you for the way you have guided the Association for the past two years and for the sacrifice you made in taking it for another year. It is almost too much to ask of a man these times, but with the present V.P. in the regular line of succession I couldn't see any future for the Association unless you did continue. . . .
> I just couldn't see the Presidency going to one of the old line hard-shell Republican isolationists. . . . They are all buttoned up nationally and should be locally.

In the early 1960s Harry served as president of the Payette National Forest Woolgrowers' Association and on the executive committee for the National Wool Growers.

Following the passage of the Taylor Grazing Act (1934), Harry served on the State Advisory Board for the Bureau of Land Management, was a member of the BLM Owyhee District #1, and

---

*The spelling of dialect is always a problem, and intimate, familiar, personal names (nicknames) are an example. Scots seem to prefer the "ie" as a diminutive, and I have seen this name spelled "Dockie." Oral tradition gives no preference. Author's note.

was selected to serve on a working committee to establish county goals for feed and grain production in 1943 during World War II.

Harry began his director's term for the Idaho State Chamber of Commerce in 1953, serving as president in 1958. In recognition of his service the Chamber named him Distinguished Citizen of Idaho in 1969, an award presented on the basis of the recipient's "significant contributions to the general welfare of Idaho through his activities as a representative of agriculture, commerce, or industry and exceptional leadership in his particular field of endeavor."

Harry was a director of the Idaho Livestock Production Credit Association for twenty-four years from 1941 to 1965. He served on the board of directors for the Idaho Power Company from 1954 to 1971 and served on the board of directors of Continental Life Insurance Company from May of 1963 to September 1972. He was also a member of the board of directors of the Shore Lodge in McCall.

Harry was selected as a member of the Arid Club of Boise in 1938, and belonged to the El Korah Temple of the Shrine, Boise; the Scottish Rites Bodies, the Masonic Lodge, Dillon, Montana; and the BPOE No. 1683, Weiser. He was selected as *The Statesman*'s distinguished citizen in the Sunday, 12 December 1971, edition of that Boise newspaper.

In 1966 Harry and Beulah donated money to the Presbyterian church in Moscow to furnish a primary room in memory of Harry's parents, Philip and Hendrika Soulen.

Harry was despondent when, in 1960, he had to have an eye removed. His spirits lifted when his old friend Dockey Bettis visited him, saying, in effect, there's nothing to it: just get an artificial eye and go on. Dockey made his point by tossing his own artificial eye onto the table. Harry's response can be charted by the continued growth of the Soulen Livestock Company, which in 1969 took in 32,000 acres of land in five southwestern Idaho counties, plus grazing land leased from state and federal agencies. The operation supported more than 10,000 sheep and as many as 400 head of cattle. The two home ranches produced about 1,200 tons of alfalfa and grains for winter feeding.

He enjoyed golf and bridge, loved shrimp cocktails, disliked lamb that had been frozen, and felt that 12 August was the warmest day for the water in the lake at their summer home in McCall, Idaho.

Harry's grandchildren recall "cocktail hour" at 5:30 when they would have Seven-Up with him and watch the news on television. Trips to the post office were anticipated outings: "Box 827, third row from the top, third box." Another favorite activity was to drive to the railroad station and count the cars on the passing trains. After reaching a final count, grandpa and grandchildren would go into the station to find out how many cars had actually passed. Report card day was always important, and "A" grades were rewarded with a dollar each.

Harry died on 1 April 1973, at the age of seventy-nine, and was buried in the Hillcrest Cemetery, Weiser, Idaho. Memorials were made to the Shriner's Crippled Children's Hospital and the Weiser Hospital.

Chapter One

which discusses transhumance and tradition and establishes the pattern of theme and variation...

It is only fair to advise readers of the scope and intent of this work at the outset so that the idly curious may be confirmed in their curiosity and the serious may understand that they, too, are justified in choosing to read it. This book, whose title comes from one of Bach's magnificent songs, is a study of context. Specifically, it is an examination of the context of a contemporary sheep outfit and of the traditions operating, in this instance, over three generations of an American livestock family with the name of Soulen. It is not an economic study, nor is it any kind of special pleading, although implicit in the work is the conviction that tradition is important. It is not a history of sheep raising, although behind the book is a pastoral activity extending back for thousands of years.

When the shepherd Corin, in *As You Like It* (act 3, scene 2), contrasts courtly behavior with the pastoral life he knows, Shakespeare raises serious questions about courtly values and about literary realism. Manners like hand-kissing, appropriate in court where hands are clean and sweet smelling, would be inappropriate if courtiers were shepherds because the hands of shepherds are greasy from the pelts, calloused from hard work, and tar-covered from the surgery on sheep. Corin concludes his argument with a statement that serves as a touchstone to understanding the values that have motivated yeomen animal tenders for generations: "Sir, I am a true laborer: I earn that I eat, get that I wear, owe no man hate, envy no man's happiness, glad of other men's good, content with my harm, and the greatest of my pride is to see my ewes graze and my lambs suck." It is a powerful statement about hard work, dirt, and what some people would call job satisfaction or psychic income. The perspective on animal husbandry is not romanticized, but it is a vision that allows both financial reward and emotional satisfaction.

It is precisely that vision which has sustained the seminomadic sheep-raising enterprise in the American West, and it is in that tradition of transhumant sheep operations that this family belongs. Not all ovine operations involve the movement of bands of sheep from grazing area to grazing area with human translocation required with each move. Farm flocks obviously do not. But it is in the very heart of processes of the true transhumant activity that tradition may be seen to operate with great vitality. Some things are done now pretty much as they have been for who can say how long. It is upon this truism that the present study rests.

A compelling paradox about this entire subject of tradition and the livestock industry comes into focus in a superb television documentary written by Steve Robbins and appearing on PBS, entitled "Back at the Ranch." Photographed by the University of Nebraska and hosted by Richard Farnsworth, the program—in a way that might not have been apparent in the beginning—presents this paradox somewhat disarmingly as Farnsworth and various ranchers comment upon ranching as a way of life and thereby contradict certain speakers for the industry, including Clayton Yeutter, whose comments generally follow the line that ranchers must become business people and that the "bottom line is... reality has changed." Reality has changed: computers must be used, artificial insemination must be employed, and embryo implants signify the wave of the future. Indeed, this tends to be the message of the intellectual establishment, from agricultural and animal husbandry colleges through the county-agent graduates of these colleges and the accountants whose services are required by the ranchers. Change must come; finding a niche and adapting to it or within it are essential to progress. To reject this line of reasoning is to sacrifice oneself on the altar of intellectual Luddism. Although the documentary dealt with a cattle ranch, the implications extend to ranching generally, for embryo transferal is done with sheep as well as with cattle.

But as with many intellectual constructs that make bold affirmations—"reality has changed"—the above message is only partially true. The story it tells is not the whole story, and my story of tradition is an attempt to outline, if not detail, other important elements. My approach to the subject of tradition in a contemporary sheep-ranching operation is that of a student of folklife. One who studies folklife studies both the folklore—that is, the traditional oral expressions shared within a group—and the material or expressive culture—elements shaped by traditional means and existing in three dimensions—also shared within a group. Folklife, then, may contain the narratives and songs of a group as well as its patchwork quilts and colored cattails. (Specific examples of verbal elements and material items are too numerous to itemize here. A helpful listing introduces Alan Dundes's *The Study of Folklore* [Englewood, N.J.: Prentice Hall, 1965].)

Tradition figures in almost all attempts to define folklife, and tradition, too, is paradoxical. On the one hand, students of folklife talk about the constancy of tradition; indeed, if one cannot detect and articulate what it is that has been passed along or handed down by imitation and word of mouth, one is not dealing with either folklore or material tradition. It is not folklore. On the other hand, folklife scholars nearly always talk about variation, about an element of change in the process of transmission that is often called dynamic variation. (Barre Toelken's *The Dynamics of Folklore* [Boston: Houghton Mifflin, 1979], offers as fine a discussion of this paradox as I know.) Engraft this understanding of paradox onto the process of sheep raising, and the partial truth of the claims of agricultural technocracy emerges. If the Soulen sheep operation is transhumance (it is) and if that operation has significant traditions embedded within it (it does), then, in the nature of things, dynamic variability will occur. Reality has not changed, but specific adaptations within tradition by which or through which the challenges of reality may best be met will occur. Tradition is thus dynamic, not static. But it is still tradition.

Bands of sheep turn out in the spring to forage the new spring growth, moving to higher elevations as grasses and shrubs in the lower elevations mature. In the fall, after summering on high mountain ranges, the fat lambs go to market, thin ones go to the feedlot to fatten for market, and the ewes go to fall pasture. After the breeding season of late autumn, the ewes move to winter pasture or to winter feeding grounds. Shearing and lambing follow in the early spring, and the cycle begins again. At each stage, herders are with the sheep, relocating as the sheep move. And it has ever been thus.

What is true of the livestock operation on which this study is based is in the main, and even in many of the particulars, true of all such transhumant operations. In a sense, this study validates Robert Frost's strategy of discovering general truths through close observation of the particulars close at hand. It is a kind of inductive analysis.

Transhumance is a way of life. Because that expression is so irresponsibly appropriated by so many who would use it to their own advantage, a clarification is in order.

A way of life makes a total demand on those who live it. In Robert Frost's words, their vocation and their avocation have become one:

> But yield who will to their separation,
> My object in living is to unite
> My avocation and my vocation
> As my two eyes make one in sight.
> Only where love and need are one,
> And the work is play for mortal stakes,
> Is the deed ever really done
> For Heaven and the future's sakes.
>   from "Two Tramps in Mud Time"

It means that those who live a way of life are totally immersed in it, living in their work, but never finishing it and never reaching a point where planned hiatuses, sometimes called vacations, can be taken without profound feelings of guilt or anxiety over work undone. Law, medicine, engineering, education: these are professions. And who would accept banking as a way of life...or management? There is no need to list the jobs that are just that—jobs. Former Yale University President A. Whitney Griswold, in a superb little book called *Farming and Democracy* (New York: Harcourt Brace, 1948), insists that of all callings and occupations, agrarian pursuits are the only ones invested with the great dignity and moral weight of a *way of life*. ("Whose Home on the Range?" *Congress on Public Lands in the West: Proceedings* [Caldwell, Idaho: Snake River Regional Studies Center, The College of Idaho, 1976, 3].)

On one issue I must concede to technology. Although I do not believe that the basic reality of range sheep raising has changed, I must quickly add "yet." Of implications raised by technology, none—not one—threatens that reality quite the way genetic engineering does. The awesome aspects of recombinant gene-splicing turn sober minds to science fiction, if not fantasy. Who can predict its outcome? If—or when—wool can be produced from willow trees and roast lamb from potatoes, then the reality will have changed and transhumance will indeed disappear.

Because the way of life the book studies has contributed greatly to Western (and western) literature, to Western development, to our creature comforts and well-being, it does not miss the mark to hope that the reader may reflect further upon the Bach chorale, shamelessly appropriated for the title of this little book.

From the playground of the small rural school, remote by time and thus by distance, we could hear the sheep and see their cloud of dust before we could see them. As they came into view following the herder leading a belled sheep up the dirt road along Monroe Creek, the homogenized bleating sorted itself out, not into the fifty different sharps and flats of Hamelin Town, but into more than thousands of baas and bleats which the ears of eight country school children tried to register. Deep bass ewes tried to summon countertenor lambs; the gamut disappeared at either end of the scale, with an astonishing range through the middle. At the end of the band came a herder, probably carrying a "tin dog"* and accompanied by real ones.

---

*A tin dog consists of tin cans threaded onto a wire of sufficiently large gauge to provide a stiff backbone so that when the ends of the wire are twisted together an instrument for making noise and thus moving sheep is available to the herder.

This annual event was recorded in those same bright images with which other important occurrences in childhood were registered. Among them is a series of pictures of my mother at the wheel of a Model T with one of her sisters beside her and an indeterminate number of preschool cousins driving across the sagebrush hills toward Quaking Asp Grove, taking groceries to one of my unmarried uncles, who was out with the band of sheep my grandfather owned. There was no road, and the trails the car took led across shallow washes through which small streams ran on their way to Monroe Creek. What a pleasant picture of rustic delight—*et ego in Arcadia sum*. In springtime, soft green grass mutes the landforms of hills and buttes that in another couple of months look dry and inhospitable. By that time the dozens of miniature creeks will have dried, their rocky bottoms dull and lifeless where once the water had given each pebble color and brightness and light. Where that precious substance flowed over patches of young grass, what was green became silver, and it was easy to imagine fish making their homes high in those draws where no fish had been since remote times in geologic history when the region was under primitive seas. And I wish to point out here the fact that twice a year—in late fall and early spring—bands of Soulen sheep trudge the route by Albertson College of Idaho where they have walked for decades. So, from my boyhood country school days to my undergraduate years and through my life's work of college teaching, I have been reminded of sheep.

Thus, the context of the Soulen Livestock Company develops. Harry and Beulah Soulen of Weiser, Idaho, got into the ranching business by buying up an outfit in the 1920s. By 1937 it was a going concern, as Mr. Soulen's brief paper entitled "Range Sheep Management" makes clear. It was natural that the Soulens' son, Phil, would take over the operation someday. After spending much of his boyhood in sheep camps, absorbing the workaday knowledge of a range sheep operation, the need to buttress practical training with the latest technological and scientific training took him to the state's land-grant institution of higher education in Moscow, Idaho, where he studied animal husbandry, range management, and allied subjects. There he met and married Erlene

Clyde; they returned to Weiser to rear a family and to work into the management of the operation.

Although the Soulen Livestock Company is far-flung, with ranges near McCall, on the Snake River below the mouth of Burnt River, and on the Payette River near Letha, the home ranches are in the high hills generally east of Weiser beyond Nutmeg Mountain. Including winter pasturage, Soulen sheep can be found in eight counties—Valley, Adams, Washington, Payette, Canyon, Gem, Ada, and Elmore. In addition to sheep, Soulen raises beef cattle. It is part of the folk wisdom of much of the American West that one does not ask a rancher how much land he owns or how many head of livestock he runs. One rancher in response to such a question answered that that was just about the same as someone asking about the size of his bank account or the extent of his indebtedness. The 4 July 1985 issue of the *Idaho Farmer-Stockman* (Intermountain Edition) featured a full-size advertisement on the back page containing the following characterization of the Soulen operation: "A diversified farming and ranching operation consisting of 40,000 acres, 12,000 sheep, and 1,800 beef cows... alfalfa, oats, and wheat on 1,500 acres of irrigated cropland." It was further noted that Phil and Harry each drives 30,000 miles a year.

So the Soulen livestock operation began, and Harry Soulen in 1937 in response to a request from veterinarian E. T. Baker, set forth in clear, concise prose the way he ran his outfit. The 1937 document establishes the general pattern of the traditions that have undergone variation as son Phil succeeded father and as grandson Harry prepares for his leadership role. The document as it was sent to Dr. Baker is quoted in its entirety. It is significant in and of itself, and it also provides the entry into the cycle of the year which is reflected in the organization of this book.

RANGE SHEEP MANAGEMENT by Harry Soulen

My old friend Doc Baker has asked me to put in writing a few remarks on Range Sheep Management. I will confine these remarks to conditions as they apply in south-western Idaho. The ewes generally run in this part of Idaho are of the cross bred type, running from quarter to half bloods. They are a large type of ewe often weighing when fat more than one hundred and thirty pounds. They practically all have a white face and are by bucks of either the Panama, Romney or Corriedale breeds.

January first will find the ewe bands on the feed yard. The bands in the fall are made up on the average of about 2000 head. The owner of this band of sheep has provided himself with 600 tons of alfalfa hay which may be enough if grass is not too late coming in the spring. Two men and a good team of horses with a 16 foot rack will scatter three big loads of this hay each day to the sheep. He will try to find a feed yard with running water as close to the hay as possible. If running water cannot be found water will be pumped into troughs. Some operators will feed grain before lambing depending on the conditions of the ewes, price of grain, etc.

If the bucks were put in the ewes on September 5th you had better be prepared to take care of the lambs which will start to drop by February 1st. As a rule the bucks are left in [with] the ewes for about 35 days, but by February 25th the lambing is practically over. As each ewe lambs she is put in a small pen about four feet square under as a rule a canvas covered shed. In twenty-four hours or less depending on the necessity of making room in the shed this ewe and lamb or lambs if she has twins is put in larger pens with in the case of singles about ten head or twins about 5 head. Of course no ewe is turned out of the shed unless the lamb is well mothered. These bunches are then gradually doubled up until they end up in bands. The ewes with single lambs are usually made up into bands of around 1000 ewes with their lambs and the twins around 600. The lambs are usually not marked and branded until they are in bunches of at least half band size.

In order to grow good early lambs it is necessary in addition to large amounts of good alfalfa to feed a considerable amount of grain. The twin ewes will eat up to two pounds of oats a day and the singles are often fed a pound of corn per day. Grain in creeps is provided for the lambs when they are around two weeks old. Whole oats are a very satisfactory feed for the young lambs after they become accustomed to grain.

In this part of Idaho any time after March 15th we begin to hope that the grass will be good enough to turn out. When it is a camptender with his pack string of four mules, a bell mare and a couple of saddle horses will start out with two herders and two bands of sheep for the spring range.

Some outfits make a practice of shearing before they leave the feed yard. They will shear late in February or early in March. Most operators however shear on the range in April and May.

If the lambs have been turned off the feed yard with a good start and the spring range and the early summer range has been good and the herder has not sicked his dogs on them too much, along in June sometime or early in July a large percentage of the oldest single lambs will be ready for market. They should by this time be fat and weigh in excess of 80 pounds.

Most of the forest permits in this section specify an opening date of July 1. There on tender feed the twins and single lambs that have not been shipped do surprisingly well during good seasons. Those that do not get fat by August 15th are usually weaned and put on pasture or shipped as feeders.

After the lambs are shipped the ewes are made up into bands again of around two thousand head and branded and the bucks are put in in late August or early September. Hampshire, Suffolk and Suffolk-Hampshire cross-bred bucks are the breeds universally used in this section to produce early lambs. Generally about two bucks to a hundred ewes are used. Some sheepmen cut the bucks out every day and feed them grain, others use bucks for a short time and put in fresh bucks while others simply put them in and leave them in the band until the ewes are bucked.

We will stay on the forest until October 15th if the weather permits and feed conditions are satisfactory. After leaving the forest we trail back to the range we used in the spring or go to pasture. Occasionally we get early fall rains with the result that we have green grass in the fall. If sufficient grass grows in the fall sheep do very well; if not, on dry feed they lose a considerable portion of the fat they had on their ribs when they left the mountains.

Generally in December sometime the bands will be heading for the feed yards and we are back to January 1st where we started a short time ago.

Some of the outfits in this section practice lambing in April as well as February. The April lambing ewes will get along with a hundred tons of hay less per thousand sheep and require no grain. The lambs however are lighter and generally speaking the percentage shipped is not as large.

I believe Doc said he wanted a few remarks for a veterinary magazine so I think it would not be out of place to mention a few of our problems.

Here in southwestern Idaho we have foot rot. Until we learned how to control it we spent about all our spare time whittling on those hard old hoofs and never did clean it up. Now we make it a practice to put our bands thru the hot vitrol [sic] whenever there is the least indication that they may break [out] with it. We find it a thousand times easier to prevent rot than to cure it.

We also have more or less trouble with "stiff lambs." We have been told it was caused by an infection thru the navel. Some years this perplexing ailment causes a considerable loss. Lambs become "stiff" when they are five days old or five months old and the symptoms are identical. Lambs born in April in the sage brush and rocks under the hot sun and certainly under nature's more sanitary conditions develop identical symptoms to those dropped under the admittedly unsanitary surroundings of an early lambing camp.

Then too we have the coyotes, bear and bobcats, ten percent of our sheep lie down and die or stray off during a year's time, the sportsmen want our summer range for the deer and elk and our water for the ducks but it's a great game and we like it.

Conditions of weather and market, lambing success, the price of hay—all these variables cause the sheepman to vary certain practices within the cycle of the year. Sometime in the 1940s, Harry made arrangements to winter most of his sheep on the desert near Swan Falls, where Soulen sheep still winter. Thus, the ewe bands are no longer on the feed yard being fed 600 tons of alfalfa hay by two men driving a team of horses hitched to a sixteen-foot hayrack. Since this is desert pasture, an adequate supply of water for the sheep is an abiding concern, although sheep require less of it than cattle and can get along quite well if there is snow on the ground. Harry started his essay with the ewes in January; my field notes describe a visit to the desert pasture on 18 February 1985:

> Phil Soulen picked me up at my home in Caldwell and we drove south down 10th Avenue to State Highway 55 toward Marsing, turned left at Saxton's Fruit Stand, drove around the lake to the road past Saxton's, and headed south to Dry Lake, after a detour to Pickle Butte. After crossing Dry Lake and driving through Melba, we reached the Birds of Prey area near Swan Falls, and from there we visited three sheep camps. Phil has 10,000 sheep on winter range, flourishing on white sage (winter fat), shad scale, and cured (still standing) cheat grass. The sheep had been trailed to their winter range from various fall pastures Phil had arranged for, including Symms' orchards, a pasturage he had not tried before. It may prove beneficial to both sheepman and orchardist—good food for sheep and cleaning out vegetation from the orchard as well as tramping down rodent burrows.
>
> On late fall pasture, it is customary to plan for 10 acres a day to supply the needs for 2,000 ewes. Winter campsites are moved weekly, and the sheep range over several sections during this period of time. Although this is desert country, water need not be hauled in for the sheep if there is any snow. Otherwise, the need for water is met by a tank truck and portable troughs, moved each time the camp is re-located. The winter range is approximately 50 miles long and from 3 to 5 miles wide.

Some sheep become crippled in the winter, and there are three principal causes of this lameness: a stick or twig becoming lodged in the cleft of the cloven hoof and freezing in place, lumps of mud from slightly thawed ground balling up between the toes (in the cleft), and the traditional crippler of sheep, foot rot, a bacteria-caused ailment. The first two are usually called "bumblefoot."

The life of the herders in the winter camp is, except for such electronic technology as transistor radios, insulated coveralls and footwear (and other articles of clothing), and sleeping bags, pretty much as shepherds in a transhumant culture have lived for years: candles, wood stoves, Dutch-oven cookery, freshly slaughtered mutton, reliable dogs, and constant vigilance. Phil tries to have two men at each camp, and the sheep wagons are comfortable and adequately large for two men. The custom-made wagons are insulated and rubber-tired. Dogs sleep under the wagon on pelts and behind a windbreak. Supplies are brought in as needed by the water tanker, which runs daily, unless there is sufficient snow on the ground to offset the need for water, in which case supplies may be brought by pickup.

The wagons are (most of them) 8 x 16 and cost $1,500 apiece in 1960.

The herders are, for the most part, Peruvian.

The four (or five) bands of sheep will start to move off the range about 26 February, band one leaving on that day and stopping the first night at Pickle Butte, the second night at Lake Lowell. The second band will leave the range the following day, the third band the day after, until all four (or five) bands have left the range on or before the government deadline. The sheep will move down Indiana Avenue, across Cleveland Boulevard by Albertson College of Idaho on their way to the shearing pens at Letha, where clipping is expected to start about 10 March. Lambing, commencing in April, will be finished by May. Sheep in good condition, as winter range sheep (winter range is from the middle of December to about 1 March) are likely to be, can travel up to fifteen miles a day, with nine to fourteen being typical. Sheep become trail-wise, and herders can rely upon veterans to be able to know their way, although one herder with a bell sheep will lead and one herder will drive. The dogs control the flanks. Sheep can make two and a half miles per hour on asphalt paving. The foreign herders are arranged for through the Western Range Association, working with the Departments of Labor and Immigration.

Two final observations. Our dinner was delicious: French bread; soup of rice (probably an instant base), sliced Vienna sausage, and mutton; and a main course of mutton, onions, and potatoes cooked in a Dutch oven; plenty of coffee and store-bought cookies. The second observation concerns the divided door on one of the wagons. Phil said that it was a carry-over from the days the wagon was horse-drawn: the bottom part could be closed, keeping some wind out, while the upper part opened onto the Jacob staff (or ladder) behind which the driver stood.

Certain items in the field notes add to the context of this livestock operation. Harry Soulen got into sheep raising in the late 1920s, before the Taylor Grazing Act changed open-range grazing in remarkable ways. Indeed, it was this act that marked the end of free and unlimited access to that part of the public domain heretofore unregulated. Thus, in 1934 the age of exploitation ended, and the slow period of rehabilitating western rangeland began. There is a distinction between national forest grazing and grazing on the lands administered by the Bureau of Land Management, and although it is not the purpose of this study to detail the long history behind those differences, it is appropriate to say that the public forest lands were withdrawn from general and uncontrolled use early (about the turn of the century) and that grazing permits were established with restrictions on the numbers of domestic animals and seasons of the year the forest could be used. But exploitation of the "left over" lands, the drier, harsher sagebrush-grasslands, continued a generation or more after the care of the forests was placed in professional hands. The Department of Interior, not the Department of Agriculture, became the entity responsible for these dry western lands, the Interior's Bureau of Land Management looking after them.

Until 1934, "coyote" or "tramp" sheepmen from as far away as California could and did compete with regional Snake River Basin growers. It was largely this conscienceless overgrazing that created the serious problems of range management that the BLM is fighting today. And it was these invasions by coyote outfits that generated most of the conflicts between cattlemen and sheepmen and even between local and exotic sheep outfits.

An observation seems apposite here. One man's lifetime is usually the poorest of yardsticks by which cultural change can be measured. After all, what is a lifetime compared to a deathtime—all that preceded and will follow the life of one person? Still, when I reflect that within my life the miles of Idaho's paved highways jumped from a few score—one? two?—to thousands, that antibiotics displaced calomel in a physician's *materia medica*, that Mount Everest was climbed, that the mile was run under four minutes, that extraterrestrial navigation occurred, that the atom was split, then I am almost convinced that sometimes a life can be useful in calling attention to change. Two further aspects of cultural change should be cited. When I was growing up along Monroe Creek, both salmon and steelhead trout used the stream. On one occasion that I recall, although family tradition holds that it happened twice, a beautiful trout with steel-blue-green back and brilliant pink sides somehow got disoriented and entered our irrigation ditch. I recall my mother and father scraping off the scales as they prepared it for the skillet. Old-timers talked about chinook salmon that had died after spawning, providing a feast for the coyotes, which, after surfeiting upon salmon carcasses, began to lose their hair. When I shared this traditional item with Dr. Lyle Stanford, a colleague in the biology department, he suggested that although the carcasses would satisfy the coyotes' appetites the food value of spawned-out carcasses would be deficient in nutrition and might well lead to hair loss. Many downstream dams on the Snake River, built without provisions for anadromous fish, have since effectively stopped the passage of those steelhead and salmon up many little creeks, each of which sustained its modest fishery.

The other aspect of cultural change to have occurred within my lifetime and one that requires a far subtler mind than mine for an adequate treatment is our shift from a nation inhospitable to the creation of a large and permanent military establishment to a nation no longer confident that legitimate needs for self-protection could best be met by civilian conscripts and volunteers to be trained and directed by a small group of professionals until the war was won. With victory, the army was to be discharged, civilians returning to their jobs, and plans interrupted by the wretched necessity to go to war for a while. When one reflects upon the cultural makeup—ethnic, linguistic, religious—of the streams of emigrants of which this nation is composed, it is by no means difficult to recall that they had wearied of standing armies, of conscription, of generations of petty and proud kings and bright military uniforms despoiling the country and generating contempt for human life. One need mention only the Thirty Years' War as a case in point. So the Mennonites and Hutterites and Quakers and Moravians (the list could be lengthened and would include even a strand within the Methodist communion) and peasants and yeomen, irrespective of religious convictions, who always seemed to bear the ultimate costs of war, weary of warfare and distrustful of a military presence that contributed nothing to the improvement of society, helped shape an American mentality that was decidedly antimilitaristic. One of our earliest traditional songs ("Yankee Doodle") comments sharply upon the wastefulness necessarily part of military establishments: Father and I went down to camp/ Along with Captain David/And what they wasted every day/I wish it could be saved.

The view we on Monroe Creek had of the regular military enlisted man was, I suspect, the ordinary American view. It was the view expressed in the remarkable social document *And Ladies of the Club*, by Helen Hooven Santmeyer: "He's been in the Army all these years, and that's as far as he's got: chauffeur to a general. Looks as if he was a hard drinker... tough and ugly, like most of the Regular Army enlisted men." (Columbus: The Ohio State University Press, 1982, 1,322.) James Jones's *From Here to Eternity* is a scathing indictment of the pre–World War II military establishment, very nearly from cover to cover. But all that began to change during and after World War II, when military careers became more attractive to a higher sort of intelligence. A new academy near Colorado Springs advanced the cause of such careers, and now our defense budget takes such a huge proportion of the entire national budget that we can only wonder what the antimilitary founding fathers would say. What the presence of women at the front will ultimately mean both to the military and to women must be decided by sharper minds than mine.

So 1934 marked a significant change in the lives of rural western people and in the attitudes toward the land as the time of open range with free and unlimited access to it ended. Fenced grazing units began to appear as the BLM did its work. Overgrazing, mainly by the tramp operators, had made the land vulnerable to invasions by exotic plants, particularly a grass known locally as stickmouth or cheatgrass (*Bromus tectorum*). That invasion happened, according to some old-timers, as early as 1914. Certainly by the 1920s lands that had supported various of the native bunchgrasses were completely dominated by cheatgrass. The take-over was the ruination of grazing, thought many livestockmen, but although cheatgrass is in many ways an undesirable annual of brief life, it nevertheless provides good early forage. When it matures, however, its seeds (awns) can work great distress in animals' ears, eyes, mouths, and throats, for the seeds have tiny barbs not unlike those on porcupine quills that constantly force the sharply pointed seeds inward. Lumpjaw is an ailment caused by such a seed working its way into the soft tissue of the mouth or throat of a grazing animal. Not only that, but stands of dry cheatgrass also create an explosive fuel for range fires. But compared to medusahead rye, which I first heard stockmen discussing in the early 1950s, cheatgrass is highly desirable. It remains to be seen whether medusahead rye will be found useful and, if not, whether the range can be reclaimed from it and rehabilitated with native grasses like bluebunch wheatgrass and Idaho fescue and such shrubs as white sage (winterfat), the two kinds of saltbush, and shadscale.

That Soulen sheep have used the same winter range for so many years indicates two important elements about the context of this sheep outfit. The first is that the grazing area is a traditional source of winter feed, used year after year. The second is that this area is not overgrazed. The BLM monitors its condition, and Phil Soulen sees to it that these sheep move regularly and do not over-graze and abuse the forage, an important component of which is white sage. Perhaps the two considerations are really one, an equable and responsible connection between a resource and its use.

Winter on the desert range. Phil Soulen and herder at the entrance to a sheepherder's wagon.

Milk cans of water, sack of dog food, and a windbreak for protection of the dogs. Winter range.

Another sheepherder's wagon on the winter range. Fresh meat hanging beside the door, wood, water, dog food, and a windbreak for the dogs indicate the self-sufficiency of this insulated portable home.

Close-up shows the traditional double door, which can be adjusted for fresh air, keeping overly friendly dogs out when the bottom half is closed.

On the desert, water is hauled to bands of sheep when there is no snow.

These troughs are filled by the tanker as necessary.

RIGHT: *This sheephook leaning against one of the Soulen sheep wagons on the desert winter range is the traditional implement for the Snake River Region. It is not the shepherd's crook of the Near East and Europe, but is designed to catch a sheep by the leg, not by the neck. The Snake River Region, actually a subregion of the American West, includes approximately 110,000 square miles, including parts of Wyoming, Utah, Nevada, most of Idaho, a good bit of eastern Oregon, and a small corner of southeastern Washington.*

When the Soulen sheep leave the winter range, the long trek takes them along a driveway established by tradition. Here they are moving north toward Caldwell, Idaho, on Indiana Boulevard.

Dogs and herders keep them moving along Indiana Boulevard. There is snow here, but none on the desert where this trek started.

The bands are trailed along the eastern side of the campus of Albertson College. Anderson Hall, the honors dormitory, sits oblivious to the movement of 11,000 head of sheep.

As they move across Cleveland Boulevard, the sheep have a good backward glance at the north side of the campus and Simplot Hall and Terteling Library. The grocery cart marks the sidewalk at the western edge of the East Cleveland Albertson supermarket.

Chapter Two

which treats of shearing, lambing, marking, and associated activities...

The next stage in the operation described by Harry Soulen is lambing, a time still full of hard work and anxiety. Mr. Soulen described shed-lambing, and although Phil has used this technique and still has lambing sheds at Letha, he chooses to lamb on the range. He has abandoned the early breeding with its February births, choosing instead to put the bucks with the ewes in November-December for April-May lambing, a more pleasant time of year for man and beast. Various illnesses are generated and compounded by the massing of animals; dispersing them onto ground purified by winter's freezing, by fresh air and sunshine, reduces health problems associated with lambing sheds.

Shearing, another element in the cycle of spring activities, often took place on the range in April or May after lambing. Since time immemorial, sheep have produced both food and fiber for human use and, over time, varieties have been developed that are tasty when marketed as lamb and which produce a high-quality wool. With the goal of maintaining that dual capability in his sheep, Phil has maintained the crossbreeding strategy which Harry described in his report to Dr. Baker: "The ewes generally run in this part of Idaho are of the cross bred type, running from quarter to half bloods.... They practically all have a white face and are by bucks of either the Panama, Romney or Corriedale breeds." On 7 May 1985 I recorded Phil as follows:

LA: Are these horned sheep?
PS: Some of them are now. I haven't had any Rambouillets for a long time, but some are horned Rambouillets, some are polls,

then I bought some Panamas which are polled and then Suffolks are polled...

LA: And you breed for both wool and meat.

PS: Uh huh. That is why we bought the Rambouillet bucks. Last two years we've used them, trying to "fine" the wool up a little, so we'll use them about two more years, crossover a single cross-back—we've got to be careful which sheep we breed them back to or we'll have a double cross-back and then we'll get a little too fine, maybe, and too tight. Then we'll quit the Rambouillets again and go back to the Columbia rather than the Panama because the Panamas... there're hardly any breeders around.

LA: So you have to be careful to get quantity of wool production as well as the quality of wool *and* all the things that go into producing a good meat animal....

Thus the types of sheep that now make up the Soulen bands differ from those of 1937, and in managing for profit—or, in these days, survival—Phil will continue the dynamic varying necessary to that end. And, thus, economics is a powerful element in the shaping of tradition, and folklorists can accept it as one of those aspects of context like, for example, the operation of community standards which help mold and adapt narrative songs to changing values.

Today, Phil's ewes are sheared prior to lambing. While awaiting the shearing, they are examined for bumblefoot and sent through a footbath to protect them against foot rot. This latter condition is a fairly common—I am tempted to say traditional—ailment of sheep. Harry spoke of a "hot vitriol" treatment; Phil puts his ewes through a footbath in a cement trough containing, among other chemicals, formaldehyde, which seems to control the disease fairly well if he is able to send the sheep through at least twice. Older traditional methods involved cutting into the hooves with a sharp pocketknife and applying various kinds of caustic antiseptics, and these are still sometimes used. I recall that one of my cousins insisted that it was all right for him and me to persist in breaking certain kinds of glass containers, thick Mentholatum and Vicks jars, because these jagged shards helped keep the hooves open and draining as my grandfather's sheep walked over them in the feed yard. I remind the reader that this method of treating foot rot is family lore passed on within the same generation, from my cousin to me. It makes a real sheepman shudder.

Since all the ewes are gathered at Letha, it is the time when all the herders and camptenders and all the sheep wagons converge. In such a gathering in 1880 or 1890, I would have heard the Appalachian dialect of Tennessee, the Virginias, or the Carolinas, or perhaps the modified Lallans of a Scottish Lowlander. Indeed, the Scots and the Irish have been identified with livestock raising since the establishment of Euroamerican culture on the western landscape, through direct emigration, through emigration by way of Canada, or through the drift westward of the Scots-Irish of western Pennsylvania or the southern mountains ("Celts and Other Folk in the Regional Livestock Industry," *Idaho Yesterdays* 28, 2 [Summer 1984]:20–29).

But generalizations are accurate only as generalizations. A listener would also have heard the hesitant English of a twenty-year-old emigrant from Smöland. My father-in-law, Magni Olafson, herded sheep when he arrived from Sweden in 1897, working for a Scots-American named Al Wilson in the same general area where the Soulen sheep await shearing. It is a fact of socioeconomic history that herding sheep was the lowest rung on the success ladder, a job that was traditionally lonely, as will be pointed out in the next chapter. It was poorly paid, although that condition was mitigated by the fact that a sheepherder was given board, room, and tobacco, a not inconsiderable compensation. In addition, to those who would endure the loneliness and poor pay, sheepherding was available to the otherwise unemployable because of a lack of job skills. Cowpunching was by contrast farther up the ladder, partly because the buccaroo had to have some equipment—a reata, chaps, a "ten dollar horse and a forty dollar saddle"—and the skills to use them.

In the 1967 movie *The Ballad of Josie*, cattleman Jason Meredith (Peter Graves) argues at great length about the capital investments, know-how, and hard work essential for success in the cattle business. But "Any idiot with a two-bit dog and a Winchester can raise sheep," he tells Josie (Doris Day). And in those pre-Taylor Grazing Act days, there was some truth in the claim, at least in respect to herding sheep as opposed to establishing and managing a sheep outfit.

But idiot or not, the sheepherder was assuredly denied the protection of any divinity thought to exercise special care over retarded people and poets. One of the dangers he faced was Rocky Mountain spotted fever, sometimes called breakbone fever for its characteristic aching agony. It was not, of course, confined to sheepherders, but the tick that carried the spotted fever organism and sheep seemed to go together. And there was no cure for it, except the victim's own stamina. Two items of traditional verbal lore that passed among regional sheep camps include the following, both related to the feared disease. In one, Dr. Jones of Jordan Valley, Oregon, enjoyed an outstanding reputation for the recovery rate of spotted fever victims he treated. According to informant Ellen Corta, his treatment was simple: never move a spotted fever victim from the place where the fever struck him. Keep him as comfortable as possible, but under no circumstances should he be moved, no matter how far from civilization his camp might be (Corta Collection, The College of Idaho Folklore Archives).

The second item is a legend about a gold deposit referred to as the Lost Sheepherder Mine. The versions I first heard are from Glen McGinnis and Frank Rambaud of Nyssa, Oregon, collected during the summer of 1954. The narratives are widely known throughout the region and have been told in sheep camps for three quarters of a century. Pierre "Pete" Rambaud was a Frenchman working as a camptender for a sheep outfit in the high butte country west and a little north of Nyssa, Oregon. In the spring of 1912, he had been in town after supplies for a herder named Victor Casmyer (or Kasmeyer, spelling uncertain), whose band was some distance from Vale, the seat of Malheur County, Oregon. When he reached the herder's tent, he found Kasmeyer delirious. Rambaud unpacked a horse, tied the sick herder in his bedroll on it, and managed to get him to a doctor in Vale, who pronounced him dying of spotted fever. After the man died and his bedroll was opened so that his personal belongings could be sent home, Rambaud found a quantity of incredibly rich gold ore (Louie W. Attebery, "Folklore of the Lower Snake River Valley: A Regional Study," Ph.D. Diss., University of Denver, 1961, 29–30).

Mention should be made, too, of Greeks, usually from the Peloponnese and, of course, from the mid-1890s Basques became more and more important. So western sheep camps knew the sounds of English intoned by a variety of linguistic types. Frank Aguirre, a Basque emigrant who came to America in 1954 and has been foreman of the Soulen sheep operation for nearly a quarter of a century, has learned the sheep business as most Basque emigrants involved in it have had to do, for tending sheep is not a customary activity in the homeland. Of all his job-related skills, none is more important than his ability to speak Spanish, for most of the Soulen herders are from either Peru or Spain.

While awaiting the arrival of the shearing crew, the herders and camptenders take care of feeding the sheep and assist in the footbath processes. When the itinerant shearers arrive and set up the portable pens and "shearing plant," the shearing itself begins if the weather cooperates by withholding rain.

Tending the sheep on the winter range, moving them in for shearing, midwifing at lambing, trailing them from the lower to the higher range—it is a familiar pattern. It is the way of transhumance; it is also in a vague and dimly perceived way reassuring. All this has been done before in similar but not identical ways. When our political parties betray us, when it appears that the small voices of citizens are not merely ignored but seem instead to be scorned, when thoughts of the potential for the dissolution of this beautiful planet oppress us—when all these considerations make us weary and fretful—the reminder that ewes will be shorn, lambs born, and stories told is reassuring. Our connection with the earth and its natural cycles is undoubtedly a necessary precondition to our mental and spiritual health. But is all this merely romantic nonsense, silly stuff that has kept the pastoral literary conventions alive since the time of Theocritus in the third century before Christ? It must be admitted that there is a semantic fracture between the words "shepherd" and "sheepherder" and that there is dirt on the hands of those who work with sheep. But human life is sustained by its dreams and visions as well as by its grasp of reality, and pastoralism in literature is not just romantic nonsense but an enduring strategy by which the imagination harmonizes our dreams and our work. It is sobering to reflect that this hardworking and worrisome way of life and the imaginative response to it in art and literature may endure only as living fossils in view of the previously cited statistics of the waning of sheep raising.

Shearing has traditionally been done by crews moving throughout the sheep-producing regions, and their lore is more than sufficient for another study. I turn to my field notes for a few details of activities at shearing time and to taped conversations with Phil for his insights:

On March 9, 1985, I drove to Letha to find out when the shearing would start and to see whether there were any other activities of a traditional nature pertinent to the study of the operation of a representative sheep ranch.

Letha is the site of shearing operations, one of the stopping points in this transhumant process that always seems to be in process.

Although the only operation that seemed to be going on was the footbath for foot rot, of which more later, there was an opportunity for surveying some of the traditional structures—sheep feeders, sheep wagons, the bunkhouse, loading chutes, and so on. One of the most interesting features of the cultural landscape was the two-woven wire fence with weed infill serving as a windbreak.

In conversing with Frank Aguirre, the foreman, I learned that the sheep are moved through the vats for foot rot treatment two or three times. This was the second treatment. The vat is made of cement; it's seventeen years old, about forty feet long, and shallow, holding just enough liquid to wet the hooves. Of interest also are the following bits of information: (1) from 1,700 to 1,800 sheep per day are sheared, depending on conditions and the number of shearers in the crew; (2) Phil raises about fifteen hogs for butchering for hams and bacon for sheep camps; (3) two bucks can service a hundred ewes if the range is somewhat confined, as in low- rather than high-country grazing; (4) 40 bucks will service 2,000 ewes, 230 for 10,000.

At 12:30 we had a good dinner of bean and meat soup, mutton stew, lettuce and onion salad, homemade bread, fruitcake, and coffee.

On March 16, I returned to Letha for photos and conversation with Phil Soulen, as the shearing process draws to a close. The twelve-man crew (at maximum strength, with a smaller number at work while I watched, since some had been and will continue to be shearing at night) rarely operated at full strength. The shearing "plant" was divided in the middle with two rows of shearers pulling sheep from an interior chute, tripping them and setting them on their rumps, shearing them, and releasing them through canvas curtains into a north or south holding pen—six curtains on each side.

The shearers don't bundle the fleece; a herder, a camptender, or other employee does that, and there is at least one per side or bank. Sheared wool is called a *fleece*, which tends to be of a piece. It stays together and is collected and baled with other fleeces. *Tags* and *bellies* (loose, belly, and other occasional clippings) are collected and baled separately.

Each bale weighs about 400 lbs.

I have some footage of Phil talking about the process.

For dinner: bread, butter, jelly, lentil-mutton soup, lettuce and onion salad, fried mutton, a gelatin dessert, coffee.

The following is a taped conversation of *16 March*:

LA: This study, as you know because I've inflicted myself on you for almost a year now, is concerned with traditional aspects of sheep raising, and there's still a lot of tradition involved in what you do. The technology will change here and there . . . the shears they're using are probably refined instruments. Certainly they aren't using the old hand blades, but a lot of this is traditional.

PS: That is for sure. Of course, they don't use the hand blades;

their equipment is much better today than it was years ago, not that a good blade-shearer couldn't shear as many sheep in a day if he kept his blades sharp as they can with this equipment, but certainly this is the more modern way of shearing, and we still have a good many people that are doing the work that are quite skilled at their job.

LA: And this is kind of a transmitted skill, isn't it? You mentioned the father and son or fathers and sons that work this crew.

PS: Yes, it is. It's something that takes a degree of time to learn and to do it properly. They have to want to do it because it is seasonal work and, of course, this particular crew... they spread their work over as much of the year as they can, with the shearing on our particular operation in the spring of the year and there are other operations that shear in the months of January, December, and then throughout the summer and fall months, while they have the tagging operations or else the shearing of lambs that are going to market—feeder lambs.

LA: You had some like that last fall, didn't you?

PS: Yes, we shear some feeder lambs.

LA: One of the aspects of studies of tradition involves the study of vocabulary used, and there's a word I want you to clarify for me, and that's the word "sloper." I heard it used and explained, but I want to get it on tape. What's a "sloper?"

PS: A sloper pelt: it's a pelt, of course, removed from the sheep when they are slaughtered, but that sloper pelt has a greater value than one that has longer wool. It's just about the right length... the length of wool on it to be used for the coats... fine coats that are made, and it comes off a younger animal that has been under good clean conditions and at the right age and right length. Now the Idaho lambs... most of the early-born lambs that go to market in this state have perhaps four-and-a-half to five months' age on them when they're slaughtered...

LA: Where does the name come from?

PS: It comes from Colorado. Lambs off the slopes in the mountains in Colorado were at that particular age and that's the phrase that was coined.

LA: Eastern slope or western slope of the Rockies?

PS: They never considered that Idaho had any of these lambs until recently, but we have the same conditions as over there.

The shearers do their work quickly. Their first strokes are long sweeps down the belly, head to tail; subsequent strokes as the sheep are turned from side to side are from tail to head. Good shearers try to keep their clippers constantly in motion and in contact with the wool. Otherwise, wasted motion and energy result in lost time. Orren McMullen was born near Drewsey, Oregon, in 1899, and when Spokane, Washington, had its world's fair in 1974, I turned to him for information on the traditions of sheepshearing since I knew he had grown up in the sheep business, and I had to line up talent for the ranch exhibits. Although he was willing to attend the fair and demonstrate the use of the traditional hand blades, his doctor would not give his consent, and I had to be content with several afternoons of conversation and with his delightful *Vanishing Kingdom* (Weiser: Idaho, 1972), the sort of book one wishes all old-timers would write and publish. I quote from chapter 9:

> When I got a little older [twelve or thirteen?], I learned to shear sheep with the blades. Most of the blade crews I knew would shear on an average of 90 to 100 a day. Some blade men would get 125; a good machine man could shear anywhere from 175 to 250 depending on the sheep. You had to be able to shear and do a good job on 100 head to get a union card. I sheared with the blades for a few years and went on to be a good machine shearer and I followed it every spring from then until the sheep were about all gone. There used to be many thousands of sheep and many shearing corrals over the country where crews sheared for a month to six weeks at a time. Our crew would start in February and we sheared the early work around home, then we would go to California for a month or two then back home for spring shearing. Then on to Montana to finish up there, usually last of June or early July. One year I sheared from February to July 17 and that year I sheared over 17,000 head of sheep. Sheep shearers were usually hard-working men and they lived hard as well. Some

were hard-drinking men and blew their money as fast as they got it. We were always playing jokes on each other in our crew. There was a bunch of us that usually always travelled together.... We got a big job away out in the hills where there had been a big corral in years gone by, but for some reason was abandoned and we were to go there and rebuild and reconstruct the corrals, buildings, and so forth. We did this and we rebuilt the old cook house. These old cook houses were just long shed-like buildings with a kitchen in one end and a long table down through the center of the building. The shearing crew would always eat first then the sheepman and his crew next. The tables would hold about 16 to 20 people (p. 65).

In describing the shearing processes of the vast McGregor operations, Dr. Alexander Campbell McGregor writes:

The arrival of... the migratory sheep shearers, in late April or early May marked the beginning of [a] busy period for the wool growers. A crew of twelve to eighteen men spent about three weeks shearing 100,000 to 123,000 pounds of wool from 11,000 or more McGregor sheep. Either hand clippers ("blades") or machine shears could be used for this arduous work. The McGregor brothers and many other Columbia Plateau sheepmen preferred to use "the blades," the traditional method. They hired unionized hand-shearing crews who had begun their work in January in the Southwest and travelled to sheep ranges in California, Oregon, Washington, Idaho, Nevada, Wyoming, and Montana as the year progressed. One highly skilled blade shearing crew, composed of Mexican laborers, and headed by Joe Lopez of San Francisco, handled the McGregor clip for more than twenty years. Machine shearing, a somewhat faster and less physically exhaustive process, was handled by several crews based on the Columbia Plateau or in nearby range states. Some of these crews were unionized; others were composed of independent "greenhorn" operators attracted to the trade by reports of the good pay earned by shearers. Both machine and blade shearers worked fast—they were paid according to the number of sheep shorn, not the total days worked. The Sheep Shearers union established a price of nine cents per head in 1905 and later devised a sliding scale of pay that varied with the price of wool. A peak of fifteen cents was reached during World War I. A competent man using blades could shear about 150 sheep per day, although an average per man was usually about 120. Machine shearers could handle up to 175 and averaged about 140 per day. The machine shears were powered by a gasoline engine. The rest of the shearing process was essentially the same for all crews (*Counting Sheep* [Seattle: University of Washington Press, 1982], 133).

McGregor describes that process as follows:

> The shearer had to grab a ewe, flip her into a sitting position, and quickly but carefully begin clipping her in long, sweeping motions from head to tail, working to remove the fleece in one piece. Sheepmen could not afford to hire inexperienced shearers who cut the fleece into ragged patches of uneven length, for woolen mills paid premium prices for uniform grades of wool. Good shearers were able to grab, shear, and release a ewe and drag another into position every four minutes (p. 134).

Examination of patents issued from 1790 to 1873 for sheep-shearing implements and machines reveals attempts to create faster ways of getting the job of shearing done. Such attempts reveal two things about the sheep raising industry. The first is the commonplace observation that in every field of endeavor there is the constant necessity to find better ways of accomplishing the task, of building better mousetraps. The second thing revealed is the fact that sheep raising reflected a general growth pattern; and as numbers of sheep and flocks increased, better, more efficient means of harvesting the wool were necessary. Thus, in April 1855, patent number 12,760 was issued to P. Lancaster, Burr Oak, Michigan, for a sheep-shearing implement; in February 1867, patent number 61,700 for a sheep-shearing instrument was issued to P. Anderson of Kalamazoo, Michigan; and in June 1868, a patent for a sheep-shearing machine (number 79,179) was issued to T. K. Alwood of Delta, Ohio. Indeed, between 1856 and 1868, thirteen patents were issued just for sheep-shearing *machines*, not counting *implements* and *instruments*. Twenty patents were assigned to sheepshearers (contrivances, not the human laborers) between 1857 and 1873 (*Subject Matter Index of Patents for Inventions Issued by the United States Patent Office from 1790 to 1873, Inclusive.* Vol. 3. Compiled and Published under the Direction of M. D. Leggett, Commissioner of Patents [Washington, D.C.: Government Printing Office, 1874], 1330). Truly workable power equipment, however, had to await the development of an efficient, portable power source: the gasoline engine.

Australian and New Zealand shearers are thought by some to be the world's best, and statistics given on NBC's *Today* show on 2 February 1987, broadcast from Australia, would seem to bear out that claim. Two shearers demonstrating their skills for the TV camera said that a sheep per minute was good time, that truly good shearers required from one to two minutes per sheep. Gary Lovell, of Lovell's Shearing outfit, which shears for Soulen, said he could do 200 lambs in a day. The men from down under are no faster than Americans, according to him. The size of the animal being shorn must be taken into consideration, and Yankee sheep are larger than Australian–New Zealand sheep at shearing time.

In the time described by McMullen and McGregor, a fleece tied by a "tier" was tramped by a hired hand into a large wool sack until the sack weighed about 400 pounds, at which time it was sewed shut and ready for shipment with the rest of the clip. Technology has varied the traditional processes somewhat. Fleeces are baled by a hydraulic press into bales, of thirty to thirty-five fleeces per bale, weighing about 400 pounds; the wool "tromper" is no longer employed here, although in Texas the tromper is still used, according to Phil Soulen. Sheepshearers are no longer exclusively men; women are part of the crew working for Soulen. Shearers are often accompanied by their families who live in the relative comfort of campers and trailers. Not all shearers eat in the common cookhouse, which, of course, technology has brought up-to-date with running water, flush toilets, electric stoves, and refrigeration. The Soulen cookhouse will easily accommodate twenty-five to thirty hands at two tables, one extending generally east-west and at right angles to a serving bar separating the dining area from the kitchen proper and the other at right angles to the first.

An important aspect of the traditions of a sheep outfit is the customary hospitality. Any visitor at a sheep camp is always invited to have a meal. Those who dispense the hospitality at the shearing headquarters at Letha include Phil's daughter, Margaret Campbell, Frank Aguirre's sister-in-law Mrs. Tony Aguirre, and sometimes their daughter Mary Ann.

Two final comments are in order before we begin to trail the shorn ewes from Letha, at an elevation of about 2,500 feet, to the lambing area east of Nutmeg Mountain, at about 3,500 feet. The first observation is that sheep are "worked" by men on foot and by dogs. That is, the sheep are driven, moved, separated into bunches, and reconstituted into bands in the time-honored way. They may be segregated by age (the short yearlings—that is, the ewe lambs about ten months old—two-year-olds, and so on), by grade of wool, by breed of sheep (the two may be the same—that is, certain breeds of sheep produce certain types of wool), or for any other reason that the sheepman might want a homogeneous band. To accomplish this process of segregation and reconstitution, bands of sheep move through a long, narrow chute with gates at various intervals leading into holding pens. The bosses, usually Phil, his son Harry, and Frank, man these gates by which sheep are shunted from the main chute into the holding pens. When a gate is opened into the pen, it closes the main chute since it swings into the traffic pattern; as soon as the gate closes, the line of sheep proceeds. It may well appear to the outsider that sheep can never be counted often enough, that sheepmen are always counting sheep. (Hence, the superb appositeness of the title of McGregor's book.) And there is more than a bit of truth to appearance, for at every opportunity counting does take place, since it is difficult to maintain an accurate count of these small gray creatures. It is a matter of economics that has generated a tradition.

The second observation is that it is with great relief the herders and camptenders gather at Letha for the various activities there. They are able to enjoy socializing with one another as the loneliness and cold of the high desert yield to comradeship and the coming of springtime warmth. And they are able to enjoy someone else's cooking. It is not that herders and camptenders cannot cook; I have never had a bad meal in a herder's tent or wagon. It is rather that someone else is doing the cooking and the dishes, and doing them very well.

The best way to describe the appearance of a newly shorn ewe is "sheepish." Suddenly, the ewe is changed from a dirty gray to almost snowy white. The rounded and comfortably plump animal looks suddenly thin, angular, and long-necked. If there is such a thing as ovine embarrassment, it is registered in those animals, which then must be branded again since the old brands have been shorn.

Conversation taped with Phil on 16 March 1985 is transcribed as follows:

PS: We hope to be through here in just a few days and the sheep will remain here on feed until the time comes when we just have to leave here and go to our lambing range and that time is nearing pretty fast—the twenty-fourth of April we will have to be pulling out of here with the first bands.

LA: Do you drive them?

PS: They will be trailed to the lambing range.

LA: What was that date again?

PS: Twenty-fourth of April. And again on the twenty-fifth, twenty-sixth, and twenty-seventh we will be leaving. That will be four bands, and probably keep the yearling ewes in and a later lambing band for a little bit longer if the hay supply holds out. It's been a long winter, and it's taken a tremendous amount of feed. That is, these sheep were wintered out, but then we plan for a certain period of time for them to be in here and staying beyond that you have to make commitments to buy more feed.

LA: You trail them from here on up to Crane Creek?

PS: Yes, that general area north of here.

LA: And that's when they lamb and where they lamb? And that's where you fight the coyotes?... They have their napkins tied and their knives and forks?

PS: Yep, they're ready for young lamb chops.

LA: I don't blame them; you can't beat lamb. Is multiple birth an indication of unusually good feed conditions at breeding time or is there any connection between the health of the animal and multiple birth?

PS: Of course there is. We have a high percentage of twinning in sheep.

LA: And you like that, don't you?

PS: Well of course. Saving them is somewhat of a problem, all of them. But it's hereditary.*... I think ours will probably have around a 150 percent lamb crop, and then it's up to us to try and save as many of those as possible.

LA: That's pretty hard to... or maybe you've never had... triplets.

PS: Oh, yes.

LA: ...pretty hard to save the odd one?

PS: The odd ones are removed from the ewes under the conditions you have here of shed lambing, and you would graft that lamb onto another ewe.

LA: Do you have to do some feeding at lambing?

PS: On the open range? No. Hopefully there's enough vegetation, but, no, strictly, when you're out, you're out. We have no roads there and they are out in rough country, and they forage.

LA: And how long before you move to the higher pastures?

PS: We leave in the month of June, early June. We leave to go to a higher elevation around 5,000 feet, this country being around 3,500 where they'll be doing the lambing, and from there, then we migrate on to the national forest which is from that level on up to 8,000 feet.

---

*Twinning is hereditary.

On 7 May, I rode with Phil to the lambing area in the hills east of Weiser, where about three weeks earlier the ewes had given birth. It was time for the activity called by tradition "marking." The name is partly accurate, for lambs are branded with the mark of the particular band to which they belong. Not only is the brand unique to the band, but the position of the brand and the color are also band specific.

On the way, I visited with Phil about lambing:

LA: Do the ewes have trouble the way first-calf heifers do? Trouble with the first lamb?

PS: Yeah, they do. They have to watch them pretty carefully. We don't use any Hampshire rams, for the Hampshire ram is bigger-headed, and people have to watch when they use those, more so than some of the other breeds.

LA: When you graft them . . . a lamb onto a ewe that's lost hers, do you have to skin the dead one . . . ?

PS: We do that often. They skin the dead one more so down at the shed than they do out on the range, but they do that out on the range, too . . . slip that jacket right over and use some of the afterbirth on the lamb they're grafting on. This has to be done early, before blow flies are out.

LA: It's an identification by smell, mostly?

PS: Yeah, mostly by smell. Any of the natural scent from the dead lamb.

Actually, there are four distinct processes subsumed under the term "marking": the branding or marking itself, earmarking, docking, and castrating, all done as steps in the single operation. Field notes of 7 May 1985 are appropriate here:

I met Phil at the Park Street Market, Weiser, Idaho, at 5:30 this morning, leaving my car and riding for the rest of the day with him in his pickup. As we traveled slowly to the first of his ranches, we visited about certain aspects of his livestock operation (principally sheep), and I put some of the conversation on tape. We drove slowly because he was hauling a horse that had recently had surgery on a leg.

After unloading the horse at the ranch, we made better time and reached the first work area about 7:00 A.M. As we awaited the arrival of the first half-band, Phil and Dan, an employee, played a hand of gin rummy. Each had brought a deck of cards. I declined an invitation to play and watched Dan eliminate with one hand a $4.00 debt incurred in an earlier game. Phil treated me to a roast-beef-sandwich breakfast he had been so thoughtful as to provide. Then we heard the sheep coming, and our breakfast and the cards were quickly put away.

The sheep were driven into a large holding pen and from there into a smaller pen in bunches from which the lambs were caught and presented for the process called marking, but which included more than just the branding. The remarks and descriptions that follow will be somewhat casual, at least with respect to organization. They are unified mainly by the setting and the work itself.

The first observation, for instance, harks back to last fall when I observed what reminded me of the two handles by which the scroll containing the Torah is unrolled and rolled up again. I refer to a long—perhaps thirty feet—strip of canvas, perhaps two and a half feet wide, used to drive and otherwise manage the movement of the sheep. It is a portable fence, the ends managed by two men with one or more toward the middle. The thought crossed my mind that the ancient Hebrews might have used such a device in controlling their flocks, a form that might subsequently have been used for the Torah. Phil said he had it made to his specifications. I must find out more about this. Where did he get the idea? Do other sheepmen use it? Might it be an example of polygenesis? Of monogenesis and diffusion?

Each band of sheep—ewes and lambs—is stamped with a mark unique to it with respect to position and color. The mark used for this band (remember that half a band was driven to this place—Upper Ranch—and the other half to a similar set of corrals near the headwaters of Little Willow Creek) is a small diamond double "X" for the lambs, a larger one for the ewes, in red stamped across the back just forward of the hips. The dye is an improvement over the old standard black dye made of lamp black and linseed oil (boiled), for this dye can be washed out of clipped wool, retains its color for a long period of time, and is easy to apply. It is an Australian product, manufactured in the United States (Ausimark trademark).

The contamination of wool is a continuing problem for sheepmen, for any stain or discoloration that is permanent will decrease the value of the clip. Black plastic baling twine is particularly bothersome, for fragments of it tend to stay in the wool, and in processes in which heat is applied, the plastic melts and stays, showing up later in the cloth.

One of the most interesting aspects of the entire process referred to by the general term "marking" is the traditional way of doing the actual castrating. As is also true of the traditional Lappish method of castrating reindeer, these sheepmen use their teeth. The bottom half-inch or so of the scrotum is sliced off with a sharp knife, and pressure is applied with the fingers to extrude the testicles out of the scrotum as well as out of the membrane covering them. The operator then bends forward, seizes the testicles in his front teeth, and pulls his head back, thereby removing the testicles and some of the spermatic cord, including the vas deferens. He retains the testicles in his teeth while he continues the marking operation by docking the tail, tossing it into its proper pile behind him, and marking the ears. He then drops the testicles into a pail with six to eight inches of water. The herder holding the lamb carries it to the man who applies the brand to it. The lamb is then released.

The tails are partially cut and partially twisted off, so as to control the bleeding. The blood vessels are on the underside of the tail; a clean cut completely through the tail would cause much bleeding. To prevent this excess, the operator twists the tail about a half turn and the woolen side is cut, after which the tail is twisted and pulled off. The white-faced ewe lamb tails are tossed into one pile, enabling the sheepman to keep count of the ewes available for replacement purposes. All other tails go into another pile, the two piles giving a tally of all lambs. I suspect docking is an American tradition; it is not done in Scotland, England, Wales, or Ireland, at least, not ordinarily, according to my observations.

In connection with this tally, a pool is developed to which all who wish to do so put in a dollar, guessing what percent the lamb crop will be. It is expected that, since multiple births are not unusual, any lamb crop should be over 100 percent—the percentage is ordinarily between, say, 123 percent and 127 percent. I lost $2.00!

In presenting and holding the lambs for their surgery, the herder picks up a lamb, grasps the front legs one leg at a time, and brings them from the inside to the outside of the back legs, setting the lamb on its rear and presenting the scrotum and the tail for their respective surgery. Ewe lambs, of course, are merely docked, earmarked, and branded. The entire operation (earmarking, docking, castrating) takes from ten to sixteen seconds.

One of the treats this activity offers is roasted lamb's tail, tossed into the fire and cooked in the wool. The skin, covered by burned wool, is pulled off, leaving a succulent wisp of flesh and a bit of chewy cartilage.

The ears are marked by a slit and a notch to indicate age, necessary in years to come in mouthing ewes to determine age and whether the animal should be replaced.

The luncheon menu consisted of French bread, potatoes, peas, and beans fried together, lamb fries dipped in flour and fried in a Dutch oven, and coffee. They are indeed excellent.

The traditional blended with modern high tech as a helicopter landed at our nooning place. The pilot and his passenger, a government predator controller, visited with us, although they declined to eat. Another official in charge of predator control was also with us, and he did eat. As I talked with him, I learned that coyotes are the biggest threat to sheep. Another threat is the black bear when the sheep get to their high summer pasture. Bears tend to kill more than they eat and some, in the course of the depredations, become choosy in their diet, preferring perhaps only the bags of ewes.

Two closing observations: (1) Lambs are grafted onto foster mothers in the traditional ways by skinning the dead lamb of the foster mother and putting the hide or pelt onto the foundling. Surely the sense of smell is one of the important means by which ewes recognize their own; probably the quality of the bleat is another means, and perhaps some general sense of how her own offspring looks may further inform the mother. (2) Although this final observation may appear to be dwelling upon the obvious, it is a matter of some deliberation the extent to which traditional practices are reinforced by university training, for both Phil and Harry (his son) are university graduates. In one sense, this is a matter far too complicated to be resolved in this study—or resolved by it—but in another sense the study cannot ignore the problem since the essence of the project is to photograph and otherwise record the traditional aspects of sheep raising.

It must be remembered that *homo sapiens* gains knowledge in various ways—under controlled conditions called scientific experimentation, through intuitive apprehensions of truth, and by relying upon and extending through analogy truth from authority. But certainly another way by which we learn and by which knowledge and wisdom are gained is experience. Surely the empirical wisdom involved in this sheep industry is large, and it is this wisdom, transmitted in traditional ways, that must engage the attention of the folklorist.

McGregor says of this procedure:

> Marking was gruesome work. Two herders held each buck lamb in a seated position on a board several feet above the corral. A third herder then used his knife, his fingers, and his teeth to castrate the lamb in a procedure long used by western range sheep operators. The grisly ritual attracted onlookers who came to the ranch every year to observe marking and to dine on Rocky Mountain oysters (*Counting Sheep*, 133).

Although I was invited to test the metal of my own knife, as well as the mettle of my makeup by trying both in the castration of a buck lamb, I declined, not because I found the operation particularly repugnant, but because I did not want to botch an operation on my indulgent host's livestock. Those wethers would be worth important and necessary dollars in the fall, and I did not want to cause the loss of even one. I grew up in an environment of branding, castrating, and dehorning (now *there* is a grisly operation), and I have seen my father use his teeth as an extra hand, in a manner of speaking, as, for instance, when he put the lines of a team between his teeth so that he could use his hands for some other necessary task. I have done the same with the reins of a saddle horse when some task required both my hands, and it was not uncommon to see someone "ear" a horse while a bridle or halter was put in place—that is, control the horse by clamping down on the refractory animal's ear with one's teeth. Lamb testicles are aseptic; there is very little blood involved, and the teeth are surer of the grip than a set of fingers would be. It is a practical and traditional way of solving a particular problem.

At the former lambing headquarters—also where the shearing is done—the sheepherders' wagons await next winter's trip to the desert range. The tradition of the double door can be seen here at the front end. As noted in the text, a team could be driven by a camptender holding the lines through the opening provided by the top half, a consideration on a cold, windy day.

Shorn sheep are vulnerable to sudden temperature drops. In Scotland the term often used to describe the shivering of a newly shorn ewe is "yow trimle"—ewe tremble. (In Scots and other Northern English dialects a medial short "i" is usually pronounced as a "u." Hence, pity *is pronounced "putty," "hill" as "hull," and so on. The spelling of any dialect word is always problematic.) This photograph shows a fence with a weed infill, a traditional protection against the wind for shorn sheep.*

These ewes, a tiny portion of the total to be clipped—11,000 or so—await their turn at the headquarters near Letha. The shearing plant is outlined by canvas sides and roof. Electric lines extend from the generator in the truck to the clippers. The bare rafters in the background are part of the lambing sheds no longer used.

A mechanical baler has replaced the wool stomper in the Soulen operation.

TOP AND BOTTOM: *Sheepshearers work on both sides of a partition separating the plant into two equal components.*

Altered in appearance from a plump, round, gray-colored creature to a thin, bigheaded, ewe-necked embarrassment, what better word is there than "sheepish" to describe this alteration? The vulnerability to cold appears obvious.

After shearing and depending on range conditions, the ewes are trailed to the hills east of Weiser, Idaho, where they will lamb. In this 1953 photograph, Fidel Uranga assists a difficult delivery as John Aspitarte lends a hand. This is not a Soulen sheep operation, but the process is similar.

Just at sunup in the hills east of Weiser, Idaho, Phil and an employee play cards while awaiting the arrival of ewes and their lambs, which will be marked, docked, and, if male, castrated.

The marking process is now ready to begin, for the young lambs are now mature enough. Among the items required are a good knife and whetrock.

Other necessary equipment includes the branding or marking dye—Ausimark—kept warm over a small fire so that it will not congeal.

Soulen's portable fence, a canvas strip about two and a half feet wide, secured to a wooden rod at either end and rolled scroll-fashion when not in use.

The portable canvas fence in use as it is held by three men moving the sheep ahead of it.

The first station at the work area is the branding or marking station. Here Phil Soulen marks the lamb while a herder holds the next subject.

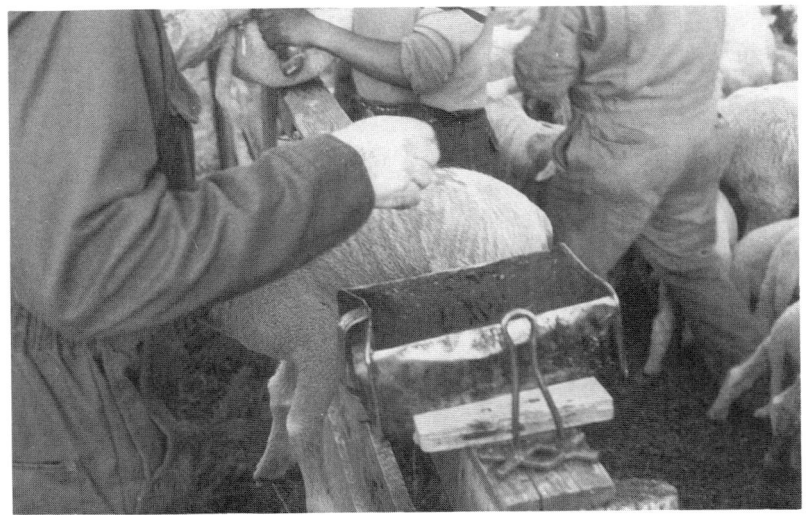

The brand is just below Phil's right hand.

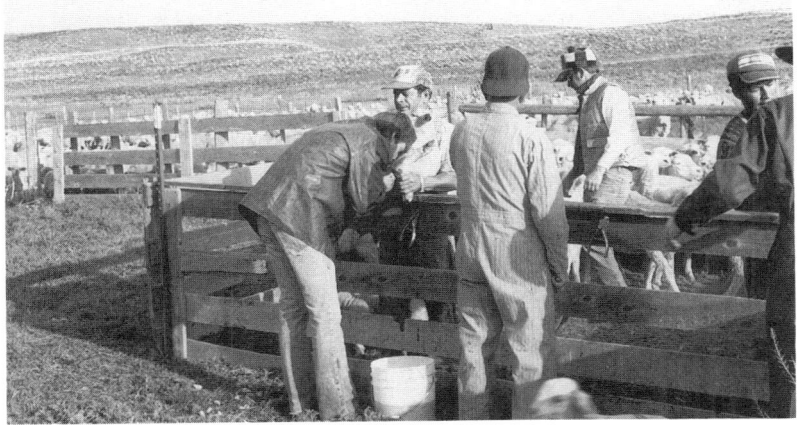

The marked lamb is then moved along either to Frank Aguirre, with the bill of his cap reversed, or to Harry (Phil's son) for docking, regardless of gender, or for docking and castrating. Harry, hatless, is bending over the operating shelf atop the panel fence. Many, if not most (indeed, if not all!), lambs in Britain are not docked.

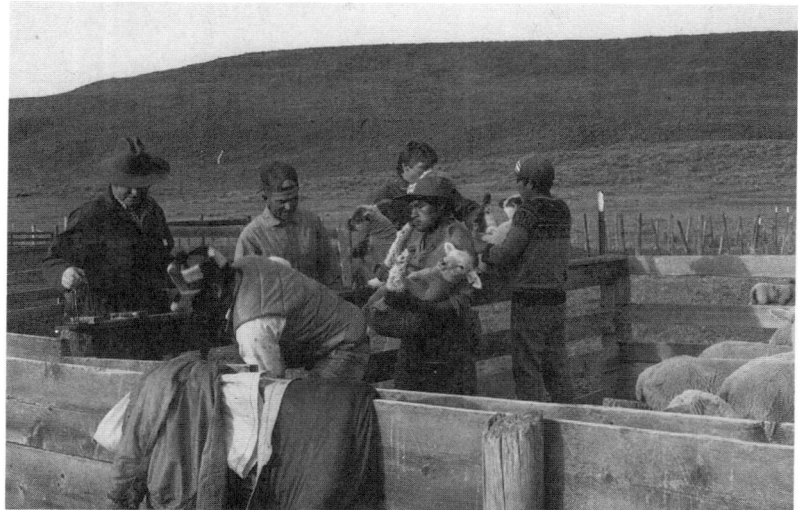

The marking-docking-castrating area, where herders or camptenders present the lambs.

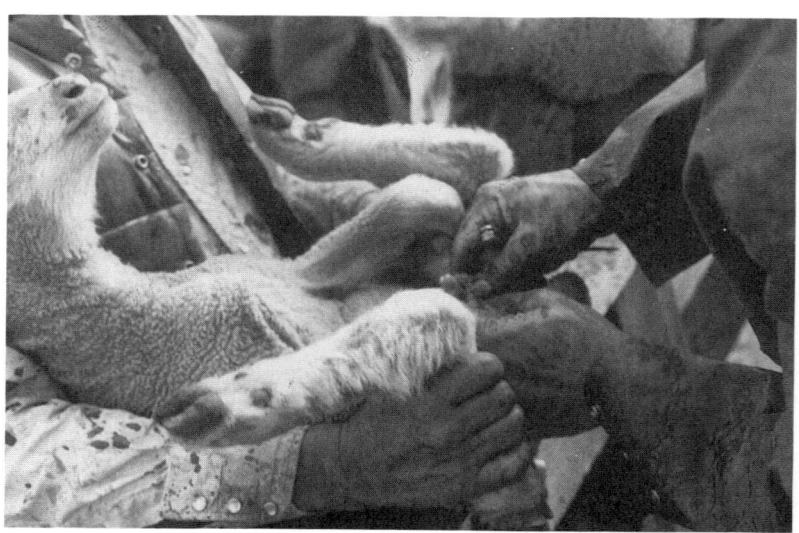

The young ram is presented to the castrator who cuts off a fraction of an inch of the scrotum, extrudes the testicles from their surrounding membrane, and prepares to remove the testicles and their attached spermatic cords.

Frank Aguirre seizes the testicles in his teeth. When he straightens up, he will pull the testicles and attached vessels (vas deferens and spermatic cords) out of the sheep. Harry has just performed this operation and holds the testicles in his teeth.

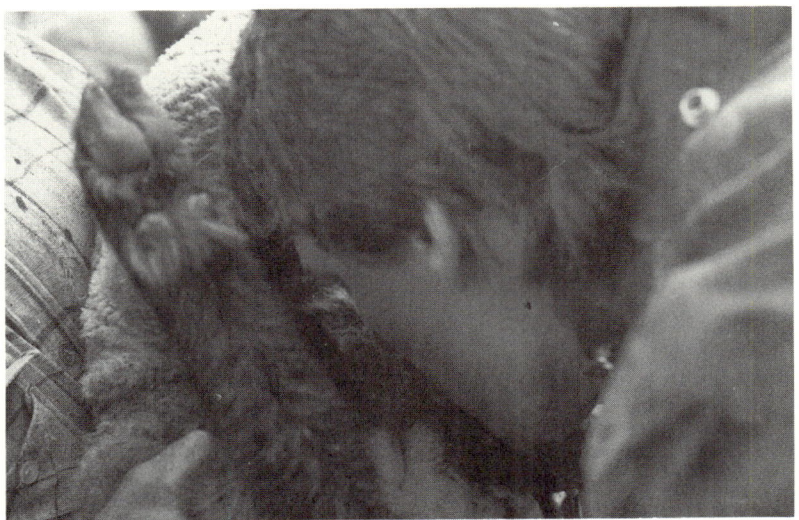

Harry Soulen, having extruded the ends of the testicles through their protective membrane, bends over to seize the testicles firmly with his teeth.

Harry then straightens up, the testicles held in his teeth, and proceeds to the next step in the operation.

Having castrated and docked the lamb, Harry bends over the plastic bucket, opens his mouth, and drops the testicles in with others. Tails accumulate in piles behind Harry and Frank.

Having just removed the testicles, Frank gives the tail a half twist, cuts it partially through, and then twists it off. All this is done according to tradition in a process that discourages bleeding. It is the final step in the process described by the dialect term "marking."

Frank docking another young lamb, the spermatic cords barely visible as they hang down the foreman's lower lip. Harry calls attention to a tail that is forked.

After a long half-day's work, Phil washes his hands in preparation for cooking the testicles at the sheep camp.

As "lamb fries," "mountain oysters," or "Rocky Mountain oysters," the testicles are deep fried in the ubiquitous and reliable Dutch oven. Drenched in flour, they fry up to a crisp and delicious golden brown.

CHAPTER THREE

which focuses upon the life of a sheepherder, how he does his work, and the provisioning processes that sustain him . . .

Harry Soulen mentioned the camptender, the pack string, dogs, and bands of sheep with their herders leaving the lambing area "anytime after March 15." It is, I suppose, the most richly evocative aspect of sheep raising: isolation, self-contained security, space, and the open air. This life is replete with tradition.

Some years ago a folklife colleague of mine at the University of Leeds suggested that we submit a proposal to the Rockefeller or Guggenheim or Carnegie or some other foundation for a grant of money by which we could undertake "a comparative study of shepherds' wagons in Eastern Europe and the American West." Stewart Sanderson had a fine idea, but we were unable to draft the proposal before he took early retirement, premature superannuation, or redundancy, or whatever the British were calling it then. I mention this because of the semantic fracture between the English word "shepherd" and the common American word "sheepherder." We would never call a sheepherder's wagon a shepherd's wagon.

Connotatively, the words are quite different, at least to the American mind, although denotatively they mean the same thing. Jesus was the Good Shepherd, the One who laid down His life for His flock. To the American mind the word "shepherd" is likely to be either entangled with Christological values and associations or overlaid with such pastoral literary images and associations as snow-white fleeces, nightingales, purling streams, cyprus trees (the evergreens of the Mediterranean, not the trees of the Everglades), and green pastures that stretch forever under a blue sky and a bright but not punishing sun. Shepherds live on milk and honey; they compose poetry and songs. Perhaps with the current fuss about cowboy poets, Bion and Moschus (among the early progenitors of the pastoral conventions) have been rediscovered not as shepherds but as buccaroos... *et* ... *Arcadia*, et cetera. In pastoral literature they have names like Thyrsis, Corydon, Adonais, Daphnis, and Lycidas, and they play pipes made of oat straw. "Oaten stops," the poet Collins called them. Pastors are shepherds. They tend the flocks in their pastorates. When exalted to positions of high rank, they even bear crooks as evidence of their pastoral obligations, stylized shepherds' crooks or crosiers.

Sheepherders, on the other hand, are like Shakespeare's Corin. They may smell a little gamey since they cannot be fussy about their linen. *Linen*, indeed! How sybaritic that sounds in connection with a sheepherder. There is not always an adequate supply of water for bathing and laundry, and their hands get dirty from handling wool and pelts and medicines and dogs. Their living is indeed close to the soil. In the previous chapter, I mentioned that their position was at the bottom of the economic and social ladder. I should say at this point that many did and indeed do climb that ladder to become independent and productive. Magni Olafson became a citizen as Mounie Olson, a successful cattleman, surviving both his study of the English language (at a Willow Creek district school as a young man of twenty) and spotted fever, although "Mounie" was too much for many neighbors who found it easier to say "Monty" or "Monnie."

How does one learn to herd sheep? In Soulen's and countless other outfits, the answer is the same as the answer to the question of how sheep dogs are taught their remarkable skills. You put the greenhorn with the old-timer, the young dog with the veteran. Then if there is an instinct or a latent know-how, the pup becomes a sheep dog, and the greenhorn becomes a herder. This traditional arrangement for the transmission of occupational skills is deeply embedded in the process of sheep raising and is wholesome even for the boss, as this fragment of conversation taped on 7 May 1985 shows:

LA: When you were a kid, did you spend time out with the bands?

PS: Oh yeah. Gosh, I spent a lot of time with the sheep. I spent time with them in the summer up in the mountains, went with them on the trails in the spring when it's really pretty... right out there with the sheepherders and camptenders, and then helped with whatever they were doing. And I went through a lambing out here in the month of April. When I was in college, why, I took a break one semester and worked through the shed lambing down below.

Taped conversation of 16 March 1985 sheds additional light on the matter of the preparation of a herder, through a knowledge and utilization of certain characteristics:

LA: We were just talking about something that's kind of interesting, and that is that sheep "buddy up."

PS: That's very true, Lou. Yearling ewes run in a band together, separate from the other ewes and lambs, and it seems that those yearling ewes will choose a partner, and so a perceptive sheepherder that is herding them will note that those two sheep are together most of the time, and if you split them up for some reason, if there is a separation, why, they don't herd well after that for a while. Pair bonding, or something like that.

So the preparation of a herder includes the development and exercise of perceptiveness, in this as in many other matters.

It may be helpful at this point to clarify a couple of terms used in range sheep operations. A *band* of sheep, for instance, usually means about 2,000 animals. However, when the pregnant ewes and unbred ten-month-old ewe lambs are trailed in from their desert winter ranges, they usually move in bands of 3,000 (2,500 ewes, 500 ten-month-olds), to be reorganized at the shearing site at Letha. When the 1,000 ewe bands bear their young, the size of the band increases by the number of lambs born. A 100 percent lamb crop would mean 2,000 head, a lot of sheep for one herder to manage. Many twin births would cause the size of the band to be cut down to a manageable number of ewes and their lambs, ideally not over 2,000 individuals.

After shearing and lambing, when the bands begin their gradual move upward to their summer ranges, each band is the responsibility of one herder. For every two bands, there is a *camptender*. Alexander McGregor quotes a 1913 study in which the duties of a camptender are outlined as follows: "[The camptender] goes ahead of the sheep, picks out a camping place, keeps the camp stocked with food and supplies, leaving the herder free to look after his sheep" (*Counting Sheep* [Seattle: University of Washington Press, 1982], 42). The camptender breaks camp, packs it on a horse or mule, and moves to the next site, where the camp is reassembled.

An important traditional activity in sheep ranching or any other transhumant system, for that matter, that has undergone only slight variation is the provisioning process. Field note entries indicate that Soulen Livestock, has, from time to time, raised hogs, slaughtering as many as fifteen hogs a year for hams and bacon for his sheep camps. Conversations I have had with Bob Skinner, a cattleman near Jordan Valley, Oregon, indicate that this practice was followed by his family, which, having arrived in the early 1860s, was one of the first to settle in the region. Hogs were raised not as a marketable commodity, but as food to be processed in traditional ways for the table—salted and smoked, not with mesquite, hickory, or some other exotic wood, but probably with alder or a variety of willow native to the region. McGregor cites an experienced sheepman named Brune remembering sheep-camp food: "The fruit was all dried—dried raisins, dried apples, dried peaches. We had very little beef..., it was mostly mutton. But we put up our own pork for ham and bacon" (*Counting Sheep*, 117–18).

Recipes for curing pork are nearly as numerous as the families who utilized this food, but one ingredient upon which all curing depended was salt, and it was employed in either a brine cure or a dry cure. The choice may have been determined by family custom, but it may also have depended on the availability of a leakproof container that would not react chemically with salt. A hardwood barrel was ideal. A brine strong enough to float a potato or an egg was prepared, and the meat was soaked from two to six or more days, depending upon the tastes and traditions of the family. To the brine could be added saltpeter, sugar, pepper, or other spices and flavorings as tradition indicated. Brown sugar was used by many families in quantities determined by custom. After the meat had soaked the proper number of days, it was placed in a smokehouse where it absorbed the smoke of alder or other wood for a number of days. Soulen had his hogs slaughtered and prepared by a custom butcher in Emmett.

Tradition directs provisioning processes in many aspects, from the purchase of goods through their preparation as table items. It is instructive to compare purchasing practices, including items themselves, of the McGregor outfit of, say, 1900, with those of the Soulen Livestock Company over the years. (See receipt dated 13 March 1900 on the John McGregor account at Gordon Brothers store for purchases for provisioning a lambing camp.)

For many years, the Soulens have bought their supplies from the same store—the Midvale Market in Midvale, Idaho. The purchasing practice is in itself traditional as well as fiscal. The store has been able to supply the various needs of a sheep operator in a reliable and consistent way at costs that are acceptable to both buyer and seller. A look at some sale receipts shows both continuity and change in the tastes and needs of men in sheep camps.

Bob Graham of the Midvale Market shed considerable light on the subject from his perspective as proprietor of the village grocery store that has supplied the Soulen operation with groceries since the mid-1930s. In an interview of 5 April 1988, Mr. Graham made it clear that he and his wife and coproprietor, Ida, operate their market within a tradition of mutual respect and trust with the Soulens. The Grahams analyze their records, note the quantities of items ordered and consumed each year, and scale the new orders up or down in response to that record. Moreover, by anticipating the needs for each spring, the Grahams watch the wholesale markets for bargains through quantity buying, passing the savings on to the Soulens. Over the years the relationship between the storekeepers and the Soulens has become one of personal friendship as

Mch. 13, 1900

John McGragor
To Mdse

Item	Amt	Item	Amt
Card seed		4 Can Corn	10 20
1 ½ ? Gloss	1 00	2 Bbl Flour	9 75
1 ?	50	2 sk Graham	1 00
sk Toothph	75	2 ca Cal Old	6 30
1 pr Shoes	3 50	5 pkg Currants	45
1 " "	3 60	5 pkg O Living	60
26¢ T Oak	1 05	8# Raisins	1 00
7 ½ Peach	8 62	Ginger	25
100 Staple	5 75	47# Br. Beans	2 35
1 Flat Hd	30	46# Wh "	2 35
3 gal Pail	1 35	6 Bh Red Bell	3 00
2 Can Tom	5 00	6 Cans Maple	9 00
3 Pr O---	1 85	1 sk Sugar	6 25
1 Bucket	45	Coffee	1 00
2 dg Pickle	2 30	Rice	2 00
1 Ketch K	60	Mace	50
2 Cof Pot	1 30	3 Bx Crackers	2 25
4 Lin Ket	2 25	2 Gal Vinig	80
2 Set K & F	1 30	16 Towels	2 80
2 Set Teas	90	Br Sugar	1 30
3 Dippers	75	5 cis Eggs	11 25
4 yd Oil Cl	1 20	10# Butter	2 00
6 pkg Cornl	60	2 Dripper	50
Walnuts	25	4 Fry Pans	1 50
After	25	2 Hats	30
2 Iron Pans	40	3 Pr Overalls	1 50
2 Lamp G	20	2 Hats	30
White Rock	20		5 00
2 Buckets	1 30		
2 Wash Bas	75		
	53 67		

Provisions for a lambing camp: John McGregor account at Gordon Brothers store in Pampa (from store ledger books of Ethel Gordon Metzger). McGregor, Counting Sheep, *113.*

MIDVALE MARKET
MIDVALE, IDAHO

1958

~~Soulen Livestock~~ — Spring order Sheep Camp

- 8 Cases Tuna fish
- 20 " String beans
- 20 " Peas
- 20 " Garbanzo beans
- 16 " Tomatos
- 9 bales 10 LB sugar
- 12 Cases gals Cooking oil
- 126 LBS bag Cookies
- 15 Cases 3 LB Coffee
- 22 Cases Peaches
- 22 " Pears
- 2 " Joy dish soap
- 6 " Chili Con Carne
- 2 " Hand soap
- 2 " stick Matches
- 20 " Chicken Veg soup
- 20 " Turkey Veg soup
- 12 " Spam
- 12 " Canned Milk
- 4 " Corn flakes
- 3 " Lge Tang
- 2 " Chore boys
- 2 " 2 LB Crackers
- 2 " Sardines oil
- 4 " Sardines Tomato sauce

- 2 Cases table salt
- 1 Case Kool aid Cans
- 14 " Vermicelli
- 2 " 47 oz Tide
- 4 " ½ gals Clorox
- 10 " ½ Vienna sausage
- 8 " dry Lentils
- 4 " apricot Pine Jam 32 oz.
- 6 " 48 oz dry Spaghette
- 10 " LB Rice
- 3 " Tomato sauce
- 13 " Toilet tissue
- 2 " 2 LB Velveeta Cheese
- 2 " Pt Mayonnaise
- 6 " Case Plumbers Candles

well as supplier-customer. As Mr. Graham remembered the magnitude of sheep raising in an earlier time, with warehouses for wool extending for considerable distances along the Pacific and Idaho Northern Railroad, serving that part of Idaho (now a branch of the Union Pacific), he also remembered some of the Scots and Scots-Americans so clearly identifiable as late nineteenth- and early twentieth-century sheepmen. He, too, recollected accounts of intemperate behavior—drinking, gambling, wenching—and mentioned one old Scot who went to a doctor complaining of something in his throat: "Doc, look down my throat." The doctor, upon scrutinizing the throat, replied, "There's nothing wrong with your throat that I can see." The Scot said, "There must be something. A herd of cattle and two bands of sheep went down there. Look again."

Some items are standard, appearing on both lists of supplies: rice, canned tomatoes, sugar, coffee, and crackers. Differences between the 1900 list and the Midvale Market list eighty-eight years later are accounted for in two ways. Changes in technology account for canned fruits and vegetables replacing dried produce, and changes in the nationality of herders and camptenders with accompanying traditional food ways account for certain other differences. Scots and Scots-American tastes ran to rolled oats (porridge), sourdough biscuits and bread (hence the "3 bbl. flour"), and brown and white beans, for these were important food items in those parts of the United States where the Scots settled (the Southern Uplands), although it would appear that western cooking traditions modified the scone-baking-powder biscuit tradition of the Southern mountaineer. The food ways of the Peruvian herders call for garbanzos, lentils, tuna fish, cooking oil, spaghetti, and vermicelli. This contemporary sheep outfit no longer packs coal oil into camps but supplies candles for light. Corn flakes have replaced porridge, and although canned milk does not appear on the 1900 list, it was widely appreciated then as it is now. Soap was obviously used then as it is now.

On the winter range, where tents would be inadequate shelter for the herders and camptenders, sheep wagons are used. But when the shorn sheep leave Letha, pack animals carry tents and supplies, and for the next seven months or so, herders and camptenders live in tents. I asked Frank how it was that a band came to be the responsibility of only one herder and whether, in truth, this tradition of one herder-one band and two bands-one camptender was wise and justifiable. His answer, which I paraphrase, was unequivocal: two herders would spend too much of their time talking to each other, and the sheep would be neglected and would scatter. One herder gives the band the attention required. Two herders sharing a camptender have their provisioning needs met adequately, they have company for a brief time, and through their camptender their contacts with the folks at home through letters and messages are kept up. A camptender will cook and work around the camp while he is with each herder. If he has a specialty, like baking sourdough bread, he will prepare his food in sufficient quantity to serve the herder in the camptender's absence. It is part of Frank's task as foreman to take supplies to the herders if their camps are accessible to four-wheel-drive pickup trucks. If they are not accessible, as is altogether likely in the higher summer ranges, then Frank will meet the camptender at a prearranged spot and unload the week's supplies—sometimes the supplies are calculated to last from ten days to two weeks—which the camptender will load onto his pack string and take to the isolated sheep camp.

If there is a typical day in the life of a herder, it might be described shallowly somewhat like this. He will leave his sleeping bag (his bed is likely no longer a few blankets and a quilt or maybe a soogan—that legendary piece of cover so nearly square that it drove the sheepherder crazy trying to make his bed—within an outer shell of canvas tarpaulin) at the crack of dawn to see that the sheep don't scatter, for they begin to graze at first light and, although their herding instinct is strong, they can scatter widely. The herder will probably have fortified himself with a sandwich before leaving his tent, knowing that he will be out with the band until midmorning, say about 10:00, at which time the sheep begin to lie down in shade, if they can find it, chew their cuds, and doze until midafternoon, perhaps around 3:00. During this time, the herder can return to his tent, fix himself a proper meal, and see that his dogs are fed and watered, if need-be. If the camptender is with him, his noon meal will have been prepared. A herder will return to his bedded-down sheep before they begin their afternoon foraging, and will, depending upon how far from camp they have been grazing, gradually begin moving them back toward camp so that they are back near the herder's camp by sundown. The herder may feed his dogs again and fix his supper. Before he goes to bed he will probably fix something to eat or at least to take with him the first thing the next morning. He will probably count his markers—one black sheep for every 200 ewes—at least once a day. When he sleeps, he will sleep with an ear cocked for predators, for storms, or for anything that might disturb the band.

McGregor, citing various primary and secondary sources, concludes that a herder's duties in the 1890s were likely to be involved with either "loose herding" or with "regular grazing." This means that the herder let his animals feed over ungrazed territory every day in a leisurely and casual manner, just so long as the animals did not separate themselves permanently from the main band. When the sheep were moved, he preceded them, slowing down the hasty, then moving 'round the flanks to control the expansion of the band, and drifting to the rear of the band to prod the languid but allowing the band to graze freely within this pattern (*Counting Sheep*, 41). Under conditions of plentiful feed, the band would "settle down to regular grazing habits," under which conditions the herder's job was "easy.... They go out about sunrise and then about ten o'clock the sheep will bed down and won't start to eat again until three o'clock when it gets a little cooler." The herder would then encourage the sheep to graze in the direction of his camp, which he would try to reach about sundown, at which time he would fix his supper, do what he could toward preparing his breakfast and lunch for the next day, and tend to his dogs. Next morning at dawn he prepares his coffee and meat and has finished breakfast in time to feed his dogs and prepare his lunch before the sheep begin to leave their bedding ground (pp. 41–42).

A 16 March 1985 taped interview with Phil on the subject of sheep losses and dangers the herder must watch for shows that coyotes are the biggest continuing menace:

LA: In addition to coyotes, what are some of the other traditional sorts of enemies or negative forces that you have to cope with?

PS: Well, disease is always one. We have very small amounts of that with these range sheep. They're on clean ground most of the time and away from the exposure from other animals, and so we have very little disease problems. The natural enemy, the coyote, is the worst, and in the high country we have the bear which marauds on occasion, and they're quite controllable [through live trapping and removal to other areas of the forest], so I don't consider them the predator that we do the coyote. The bobcat on occasion kills some lambs, and eagles will kill young lambs in the spring of the year, and we have some loss from that. The raven is another one that with a newborn lamb that hasn't got up and running about, why, they'll land on them and peck their eyes out.

LA: [Citing a visit to a livestockman in Northumberland, England] That's exactly what Charles Armstrong told me. It's one of the problems in the British sheep industry.

Now the sheep... is dominated by a herding instinct. Does this ever work against the sheep? Do they stampede, or do they crowd themselves into corners or box canyons? Do they spook in lightning-thunderstorms?

PS: Yes, lightning or thunder as such doesn't seem to bother them, other than the fact that banding together when they lie down to sleep, bed down, we call it, they may cluster [under a tree] and if lightning happens to strike that particular tree, you can lose some big numbers. We lost 185 sheep on one occasion when lightning hit a tree and killed those at the base of it. [He cited sheep loss from the wind blowing a lightning-killed tree over onto a bunch of sheep bedded down under it.] We've never had the loss that some people have from a bear stampeding a bunch, maybe over a precipice or down a hill into a brush patch, like that where the alders grow thick and come to a "V," they jammed them right down in there and the sheep piled on top of one another and smothered. People have lost 500 or more that way. I lost 225 ewes on one occasion where dogs got into them and ran them over an embankment, piled them against a set of mangers and feeders there. That happened in the middle of the night. We can have some bad losses.

LA: Was that up in the forest?

PS: No, that was in Payette, Idaho, right there with domestic dogs. No, the coyote doesn't cause a piling up. Normally, they're more stealthy and quiet. They slip in and cut one off and run it away and kill it. But the bear gets in the bunch, and he just starts slashing, and after this has happened—and they come back repeatedly if you haven't eliminated him—why, pretty soon the least movement and the sheep are afraid, and they start to move and run and that causes a bad loss.

At three times in the course of the year's activities the solitary life of the herder becomes a social life—at shearing, at lambing-marking, and at fall gathering-shipping. These are occasions for the telling of stories, for sharing accounts of what has happened since the last meeting, and for pulling pranks and engaging in the horseplay of working crews of men. Although each of these deserves a full treatment, I am more concerned here with a general description than with a detailed and thick description. Traditional storytelling or narration is often misunderstood. I shall try to make clear what I mean by this process.

> Studies of folk narration (story telling) may concentrate upon the teller, upon the story told, upon the matrix within which the teller and his audience meet, upon the structure and form of the story, or upon the function or purpose story telling serves. All are legitimate subjects for study.
> It may be helpful to establish what story telling usually is not. It is not often some kindly grandmother sitting with an adoring and enraptured group of toddlers reciting "Hansel and Gretel" or "Snow White." Indeed, many of the traditional folktales called *märchen*—"Cinderella," "Puss in Boots" —were told to an audience of adults. Story telling is not confined to the chimney corner, nor does it serve to while away long winter nights. Stories told are not always clean and nice,

holding out rewards for the virtuous and punishments for the vicious. Rather, many folk narratives are biographical and autobiographical, telling about the teller or about those whom he has known. Thus, many folk narratives are true, true to the sense of remembered actuality. Obvious exceptions, of course, include the tall tales or windies, those exaggerations nearly always recognized as such except by hearers who are not part of the community. Many tellers or folk narrators are men, and they tell—or told—their stories under almost any circumstances: while working; before work; after work; during, before, and after meals; during social visits; at dances; in the pool hall; in the barbershop.

We should remember that story telling flourished, and flourishes, in a time when and in places where there were no or few communications media. We should also remember that people need entertainment. If the stories deal with the sometimes harsh practical joke, with the animal functions (flatulence is a recurring motif and sometimes it is a source of amusement, sometimes a source of embarrassment, sometimes both), with local retardates, we need not be surprised. We can see the same things on "Saturday Night Live."

Folk narration is an invaluable source of legends and memorates—the very essence of what is popularly called oral history. A good storyteller can, thus, bring to life the mood, the attitudes, the characters, and the events of a time only remembered, with many of its hopes and fears, accomplishments and failures. But women were, and are, storytellers too, and their narratives played an important role in their lives. A problem that needs to be studied is the matter of whether women tend to tell certain types of stories with some being reserved for men. It is known that in many instances a particular story will be the preserve of a specific story teller, and although the substance of the story is well known by many narrators in the community, they decline to tell it since "That's Jimmy Applegate's story, and you better get him to tell it to you." Does the same prohibition exist with respect to types of stories being the preserve of one sex or another?

Much more may be said about folk narration, but for our purposes we need to keep in mind that story telling involves much more than telling *märchen*. The traveling salesman joke is one of the best examples of the contemporary folktale (*Northwest Folklore* 6, 1[Fall 1987]: 51–52).

On 14 September 1984 at Pearl Creek Camp north of McCall, Idaho, some of the bands of sheep that had summered on the Forest Service grazing units began to gather for sorting and shipment of the fat wether lambs. It was a time for work and play, and I quote my field notes for the next three days:

14 SEPTEMBER, FRIDAY:

Arrived Pearl Creek at 5:45 P.M., from McCall, which I had reached at 4:55 P.M. Frank Aguirre drove me to the camp where two Peruvian herders and—of all people—Judd Beeson were cleaning up around camp. Of the traditional items in camp, surely the following have not changed much over the years: tents (2) 9 x 11, grub boxes 2 x 16 (deep) x 12 inches (These are two feet long, sixteen inches deep, and twelve inches wide. They were designed to accommodate two rectangular coal-oil cans. The alforcas—that is, the canvas containers into which the boxes fit, containers and contents fitting onto the packsaddle—remained the same size over the years, even though kerosene is no longer a sheep-camp item. The grub boxes wear out over the years and are replaced, but the same alforcas have been used year after year), metal camp stove, sawbuck packsaddles... I'd like to photograph and describe the grub boxes better, later. It's now 6:35, perhaps too dark for pictures and "we" are bringing the sheep down for counting. Harry walks ahead, Phil leads a belled sheep into the enclosure and the others follow—slow, deliberate human movement, for the most part—Frank just jumped and waved his arms—a prring sound, the way Glenn used to call sheep; Harry calls out "hundred" and Phil responds "hundred"—five (six counting me) moved them in for chute counting from the mountainside. Phil makes a mark for each 100.

Counting was finished at 6:50. Judd (John) Beeson speaks good Spanish and has done quite well with Quechuan.

2nd band arrived at 7:02, counting started at 7:16. Harry counts with a definite counting motion with his hand. It takes about 30–40–45 seconds for a hundred to be counted and walk by. An eight-man crew, counting me, three Peruvians, two Soulens, Beeson and Anderson—back to camp for coffee. Back to town, washed up for supper by 8:35 P.M. Phil took us all to a late dinner at a local pizza parlor.

It is now 11:25 P.M.—A wool growers-marketing official is staying with Phil—Tom Boyd—and he and Phil talked some pretty technical sheep jargon. The pelts, Mr. Boyd says, are superior and will probably sell as slopers.

16 SEPTEMBER. I failed to mention that when I arrived at the Soulen place in McCall, Phil, Harry, and Frank were discussing the next two-three-four days' strategy—up this A.M. at 4:45. Chute should have been fifteen inches wide, but was twenty-three and a half instead, so a 2 x 12 was nailed on giving an internal width of eighteen inches (almost). Began loading sheep about 7:30 A.M. after ham and eggs at camp—two triple-deck truck and trailer rigs; men use canes and short staffs but no shepherd's crooks; trucks are three-deck, trailers four-deck. The last truck was loaded and pulled out about 10:25. Remaining sheep had to be counted and sent out to another pasture at Sater Meadows—finished by 11:25—about 1,250 sheep total.

Camp #2, Salt Box. We reached Salt Box about 1:45—three tents here and another loading chute-corral. Chute is seventeen and a half feet interior width, forty-one feet high outside, about twenty paces from widest part of chute to back; fifteen paces along the fork-Y from twelve to fifteen inches. I hope photos will show what I am inadequate to draw.

The pens are sort of like four-leaf clovers, two each side for working-sorting sheep.

Supper: soup (vegetable and chicken) followed by a pan-fried omelette of diced onions, beaten eggs, dried potatoes—good. Lots of coffee drinking—to bed at 9:45.

SUNDAY, *16* SEPTEMBER. This day began at 4:30 with the camptenders getting around and building fires, making coffee—up at 5:00 with a hot cup of coffee, breakfast at 5:30—rest of crew—Frank first, then Harry and Phil and truckers—about twenty-five of us in all—Judd Beeson and wig—*la peluca*. Began working sheep about 7:15—three multi-level truck-trailer outfits and at least two small trucks—loading started about 7:30. I helped move sheep into trucks. After noon meal I helped load, and at one point kept tally—500 lambs, 1,400 ewes—helped break camp and move to Sater Meadows—rode with Harry. Then to town and left McCall at 6:00—home at 8:45. In the sheep camps, there is the sort of horseplay—even with Peruvians—one finds always with crews of men. The more herders the more horseplay. While loading "feeders" and "fats," Tom Boyd told a good story about Stringer-Bartlett sheep and Senator Stanfield: the latter was blocking the way to some public land by forbidding travel over his land (purchased school sections). The senator stood with arms folded and told Stringer "The only way you'll get through here is to run over me." Stringer let out the clutch and ran over him, backed up and ran over him again.

A band of sheep require about fifty pounds of salt every other day. Combed vs. carded wool: combed, all the fibers are parallel; carded, no effort is made to parallel fibers. Worsted cloth is made of combed wool and is superior to carded.

At Salt Box Camp, meat saw is hung on ridgepole of tent. Frying pans and kettles dry by being placed upside down on the stove. The feeders that won't be fat for forty-five days or more will be sheared at Letha. A herder may or may not go out to his sheep before *desayuno*—breakfast—depending on how far they are from camp; a long way off he will eat, then tend his sheep. A bell sheep will be followed by a band or bunch, and the bell sheep is often a pet—that is, a sheep that some youngster has prepared for show at a fair.

While working at loading sheep, Phil told a story of Sugar Mountain (a Virginia camptender) and his buddy who were mining in the area of the camptender's herds. They ran out of whiskey and headed for town in the camptender's pickup truck. On the way they encountered a bear somewhere between Squaw Meadows and McCall. An axe was their only weapon, but the wielder was slightly off the mark, and the bear climbed the tree with the axe in his skull. After some deliberation, they made the long drive back to the sheep camp, got a rifle, drove back to the treed bear, shot it, and loaded the carcass into the pickup, the axe firmly embedded in the bear's skull. In town they celebrated for two or three days while telling bear stories, and the bear got riper and riper. One herder had a glass eye which he would remove to get attention or to make a point. By the time the herders had finished their drunk, it was a toss-up as to whether they or the bear smelled worse.

NOTE: Sheep moving in the wooden chutes do go "trip, trip, trip," just like the Billy Goats Gruff.

Some comments about the function of these narratives is appropriate here, and in no particular order they are as follows. These narratives are esoteric—that is, unique with and peculiar to sheepmen—owners and herders. The stories reinforce their traditional values. They validate and clarify the postures and strategies through which the sheepmen meet the world. These narratives are the verbal means through which a world view is articulated. It is consistent with the nature of esoteric lore that to tell these tales is to be identified as a tradition-bearer and a value-shaper, for these tales remind the teller and inform or remind (or both) the audience of the early times of sheep raising, what its origins are, where it came from. The journey from that point of origin has witnessed independence, loneliness, violence, hard times, and the validation of traditional ways of doing things.

The Stanfield-Stringer episode is a narrative that is humorous, whether the event itself was funny or not. It is an account that allows the teller to emphasize, if not exaggerate, Stanfield's self-important obstructionism: what right had he, a United States senator, to keep a little man, a yeoman sheepraiser, from pasture that belonged to the latter? At the same time, the narrative allows the teller to emphasize the independence and self-determined effort to achieve justice without recourse to the courts. The law exists as a last resort, and the valiant yeoman and justice are both better served through unreflective on-the-spot reaction. Other narratives of a humorous nature are grounded in Rabelaisian jokes, monumental drinking bouts, wenching, and the like. It is tradition, not chance, that led to the erection of pseudo road signs warning that sheepherders on the way to town have the right-of-way. On the surface, the Lost Sheepherder Mine legend (chapter 2) tells of isolation and vulnerability: this occupation may be hazardous to your life! But embedded in it is the motif of hidden treasure, of gold right under one's nose, not acres of it, but enough. And it vanishes before it can be claimed. It is a parable of the lost or ignored opportunities of life.

The narration of Sugar Mountain and the dead bear is a superb evocation of activities, associations, and smells that Rabelais would have understood.

Both Phil's story about the fragrant sheepherders with their smelly bear and Tom's story about the senator and the obliging sheepman were humorous and light, clearly told for entertainment, although other and perhaps more serious values are embedded in them. It is accurate to say that the tellers derived as much pleasure from the telling as did the audience. Both stories were told with no tape recorders used, and both included details that the summaries omit. They were seen to fit an occasion, they seemed spontaneous, they grew naturally within a social matrix. A year later, I was able to get Tom to retell the story of the senator, and a transcript of it follows:

LA: Tom, you've told us an excellent story about Stanfield and another great sheepman in eastern Oregon. I wonder if you'd mind sharing that with me.

TB: Certainly. I was very young when this happened. I don't think I was even in grade school yet, I remember the excitement it caused around Baker and how John Stringer came west working for Swift and I think he was from South Carolina, and later he and Dr. Bartlett from Baker formed a partnership and they bought up some school sections while Huntington Land and Livestock was possibly—well, I'm sure—one of the very biggest sheep outfits in the West.... I think they were reported with something like 120,000 ewes, and one of the principal owners was Senator Stanfield. And they didn't run on deeded land—they ran all on open free grazing—you know, just public domain. After Stringer and Bartlett formed their partnership, they started buying some school sections scattered throughout, and as I understand it most of the sections they bought were the sections that contained the water. And, so they bought these school sections right out through the middle of the Huntington Land and Livestock's land that they claimed for their grazing. And of course, in those days, open range... and I'm sure you've heard of many stories about the range wars, well this did cause some problems—so, what they would do then, they'd trail—I think Stringer and Bartlett just had at that time probably just a few bands of sheep—two or three bands, maybe. And—that they'd trail from one of their school sections to another school section, and when they were going from deeded land to deeded land, why of course they were on the open land which the Huntington Land and Livestock claimed to be theirs. And so this went on for some time and John Stringer was on his way in to tend the camps, haul groceries into their camps, and Stanfield met him out of Huntington where the road ran back into range country, and he was standing in the gate and told Stringer that he wasn't going through that gate, that he wasn't coming on into their property or into their grazing land any longer. And, this Stringer was driving an old touring car. It seems to me I was told it was a Studebaker. He told Stanfield to get out of the gate, and they say Stanfield crossed his arms, spread his legs, stood in the middle of the gate, and said, "The only way you'll get through here is to run over me." They say ol' John just put it in gear and he ran over him. And he put it in reverse and he backed into him! It sounded like he had good intentions, but I guess all he did was break his leg, so Stanfield ended up in Baker in the hospital to have his broken leg put into a cast, and he swore out a warrant against Stringer for attempted murder. So John Stringer was arrested and brought to Baker, and after a few days, why, he kinda had things organized around the city jail. He had him a telephone installed, and he had a desk moved in one cell and kinda set up office, and it was the first time I ever remember seeing John Stringer. I went there with my father and some other people that were interested in a band of yearlings... Stringer and Bartlett had a band of yearlings they were wantin' to buy, so I went with my father and we all went up to the county jail to see John Stringer. It's hard to tell how, if all this had gone to trial, why—just what might have happened, although I think Stanfield's popularity was declining, and as to what happened—well, Dr. Bartlett was a pretty shrewd person, and he kinda got to the press and here's the United States senator trying to run his livestock on public domain and prevent a small sheepman from grazing his own deeded land. So he got the press in it and they kinda got it going pretty good and they put a few phone calls and telegrams back to Washington, D.C., where the eastern papers picked it up. Stanfield started to get a lot of real bad publicity out of it, and so I think he thought the thing wasn't really working his way, so he withdrew the charges, but it really was too late—the public opinion against him would turn already pretty drastically, and that coupled with a little later story of Stanfield being arrested in Baker for being drunk and disorderly, I think about did his political career in—he was defeated and was never in the Senate again, but it was... I remember at the time there was a lot of excitement over Stringer and Stanfield, and John Stringer running over this U.S. senator that was trying to prevent him from going on his deeded land.

The fuller story is emphatic with respect to facts and to accuracy. It is the sort of narration that often occurs when an informant talks *at* a tape recorder instead of *to* friends or associates in a dynamic context in which stories are shared for the purposes of establishing a community of interest.

Journalists, faced daily or weekly with inches of white space to be filled, regularly read last year's newspapers to find out what was going on then that might help fill today's space. Every folklorist can anticipate a phone call before Halloween, Groundhog's Day, and Valentine's Day. Invariably the same questions are asked and answered as the journalist seeks a new warp for old material. Such superficial research could more easily be done in a library, but books are mute. They do not respond to telephoned requests. Hence, one can find in the earliest accounts on record here in my region the "news" that "_____, prominent sheepman of the Valley, began his shearing today and expects to have one of the largest clips he has ever shipped." Except for the magnitude of these activities (as noted earlier, fewer sheep are raised now than at earlier times), the stories of a photojournalist who covered a lambing, a shearing, or a shipping in 1897 carried essentially the same information that one finds in *Life* for 1937 or the *Idaho Statesman* for 1987. A few variations, enough to excite a folklorist and reflective of the kinds of adjustments made by economics and, to a degree, technology, can be pointed out.

For example, *Life* magazine of 24 May 1937 (2, 21: 36) notes, amid characteristically fine photographs of the Fred Holmes and Jack Wing operation based in the Sacramento Valley, that Scotsmen* and Basques predominate, that sheepherders get $60 per month and food, and that lambs are tallied by the number of tails docked at marking time. An innovation comes to the attention of those on the watch for variation: lambs were castrated in this 1937 enterprise not with knives and teeth but with a surgical tool with opposable blades, the handles of which were about a foot or a foot and a half long. This is not a process that Phil Soulen employs, the older, and in his judgment, superior teeth-and-knife technique employed instead.

As for wages and the traditional Scots-Basque employees, a feature in the 28 February 1987 issue of the *Idaho Statesman* by Charles Hillinger of the *Los Angeles Times*, introduced by a striking photograph of two herders, two dogs, and a sheepherder's wagon, carries the following information: Vincente Valle, one of the herders in the photograph, is from Peru. He earned $600 a month in

---

*In one of his continuing series on Idaho history, former head of the Idaho State Historical Society Arthur Hart devoted a column to the Scottish pioneers who helped create the wool industry in Idaho—John Hailey, descended of Scottish settlers in Virginia; John, Peter, and Hugh Fleming; James Welch and George Scott; John McMillan; Andrew Little; and the Laidlaw, Fraser, and Skillerus families (*Idaho Statesman,* 19 March 1987).

Gem County, Idaho; at home he would have been pressed to have earned $60. "It means... I have more money than [my wife and I] ever dreamed possible." Eusebio Jayo, the other herder, noted the commonplace false belief that many people assume all Basques were sheepherders in their homeland. It was a skill he learned after coming to America, as was generally true of the relatives, near and remote, of the first Basques to arrive in sheep country in the 1800s. "'It was an available job,' he said." He goes on to say, as reported by Hillinger, that when he arrived in 1969 he was paid $225 a month and given food and that now he gets $700 a month. He figures he has saved enough money to return to Spain and retire in comfort. In an interview with oral historian Madeline Buckendorf (2 October 1986), Stewart Cruickshank remembers that in the late '20s and the '30s wages got as low as $15 a month including board and room, although his dad never paid less than $30.

The feature concludes: "During the nine months they spend in the high mountains, the men sleep in tents and move from meadow to meadow on pack horses. They brave blizzards, subzero temperatures, hailstorms, thunder and lightning and heavy rain as their sheep eat 100 miles into the mountains and 100 miles back to the foothills in yearlong cycle." And it has ever been thus.

Earlier in this chapter, I mentioned the humorous signs that one can see from time to time that sheepherders going to town have the right-of-way. The signs themselves are part of the traditional humor that is directed toward sheepherders, emanating both from within and outside the group. There are many jokes about sheepherders going insane from counting sheep, from eating bad cooking, from trying to find the long axis to a soogan, from loneliness. There are stories about sheepherders going to town after their flocks have been sold and spending their wages in riotous living, indulging in the pleasures of the flesh. Such stories are not just funny, but they call attention to the harsh realities of a sheepherder's life and affirm that that life can be triumphant.

Stewart Cruickshank tells of a sheepherder who used tallow to cure his ulcers, and almost buried in his account is validation for what I said about Rabelaisian debauchery. He was interviewed by Madeline Buckendorf:

SC: He never worked in the wintertime, you know. He just spent his time in town, and he spent all his summer's wages on whiskey and women, I guess, 'cause when I got him out of there in the spring... and he worked for me for about ten, twelve years just in the summer as a herder, and he was a real good herder... when I got him out of town, out there to sheep camp, why his ulcers would be—I mean he just was dying with ulcers. And he would butcher a mutton, and just take the leaf fat, and boil it, and get that tallow out of there, and then just drink that tallow, that warm tallow, and in two or three days he was well. Cured his ulcers. Coated them over, I guess. I don't know.

MB: Either cure him or kill him, I guess.

SC: Yeah, and then he'd be in good shape 'til he got to town the next winter. But all those old herders would go out in the spring and stay 'til fall, you know. They never go to town in the summertime. Stay out there from, oh, ten, eleven months. Some of those old Bascos would... I know the one that worked in our outfit never went to town in seven years. What little stuff he needed, some clothes or tobacco, he'd just order it, and they'd take to him, shoes.

MB: It seems like it'd be a lonely life.

SC: Oh, there's people that like it. Hell, I like it myself. I think that some of the... I spent, well, right in sheep camp, close to twenty years. Never left there, that was my home. And I liked it, 'til I got married. From the time I got out of school, 'til I got married, I never... I stayed, that was my home, right there in sheep camp.

Breaking a sheep camp near McCall, Idaho, fall, after shipping the fat lambs.

*Packing groceries into a wooden box fitting within a canvas alforcas—**la bolsa con la caja de provisiones,** according to the Peruvian camptender.*

Some wooden boxes for provisions or cooking equipment tie directly onto a packsaddle.

Camptender at work, tying boxes directly to the packsaddle.

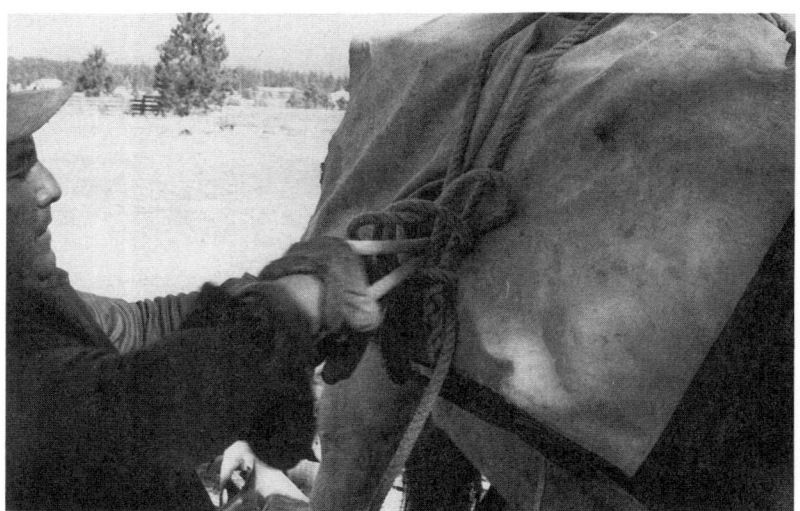

Ropes used to secure packs to packsaddles are long and the packer uses a series of loops and knots—"hitches"—in order to avoid pulling the entire length of rope through each knot.

Even the traditional sheepherder's stove is disassembled, with legs, grill, and stovepipe packed into the firebox and oven to save space. It will then be tied onto a packsaddle.

A packed animal with two canvas alforcas (one on each side) and a tarpaulin-covered tent.

The camptender leading his pack string toward the next place where he will set up camp for the herder who will follow with a band.

After grazing all spring, summer, and early fall, the sheep are gathered in the high mountains above McCall, Idaho, and sorted. This is a picture of a high mountain sheep camp. Notice the pile of saddles and packsaddles to the right of the tree.

One way of recording the results of the incessant counting of sheep is to carve a notch on a stick for each tally of 100.

Sheep are moved from the remote mountain grazing areas to corrals for sorting, counting, and shipping.

Corrals, showing gates controlling the movement of sheep for purposes of sorting.

Water for camp dipped from a clean mountain creek.

Enterprising sheep dogs drink from a pool just above the place from which water was taken for camp.

Candles are still an important item in a sheep camp. This handmade holder supports a candle inside three nails. It rests upon a metal jar lid which is fastened to the wooden holder by the three nails supporting the candle.

A sheep camp with wooden pack boxes, canvas alforcas, and various bedrolls and cooking equipment.

A sawbuck packsaddle onto which either alforcas or wooden pack boxes are tied.

At one of the mountain campsites, crude tables from earlier days make eating a bit more comfortable.

Sheepherders carve aspen trees and the tops of tables. It is a traditional declaration of identity and expression of loneliness against a physical background that tends to diminish a human being.

If the word "traditional" seems to be omnipresent, the reader is urged not to be offended. It is the key word in understanding the operation of livestock enterprises over time. Even the drying of the heavy cooking utensils is traditionally done by inverting them, washed, on top of the sheepherder's stove.

Camptenders are as clean as circumstances permit. Certainly dishwashing occurs after each meal.

Chapter Four

which touches upon the fabled hostility between sheep- and cattlemen, meanders over the problem of urban-agrarian perceptions, and concludes with some observations about reality and whether it changes...

A news item in the 25 January 1988 issue of the *Idaho Statesman* begins: "The sheep industry in Idaho is reviving after a decline that began at the end of World War II, the executive director of the Idaho Wool Growers Association [Stan Boyd] said." After mentioning increased per capita consumption of lamb and a sound wool market, the director noted an overall decline of 1,000 head in the breeding-stock population. However, the average age of the sheep was young, a sign of likely expansion ahead, he believed.

It is a precarious hope.

In worrying through what to include (which means, really, what to leave out) in this study, I spent some uncertain time over the sheep and cattle wars. Of course, they really took place. Men and animals were brutalized and sometimes killed. According to Dean Anderson of the Agricultural Research Service of the U.S. Department of Agriculture (*Science and Children*, February 1987, p. 4), more than 20,000 sheep and over thirty cattle- and sheepmen were killed in contests based upon beliefs about the incompatibility of grazing sheep and cattle and about territory. Even men within the same industry sought advantage over one another, sometimes through assault with a Studebaker or a coyote outfit, or through other means. The subject has been so debased through poor fiction and cheap movies that to include it here might invoke some kind of literary Gresham's Law. Be that as it may, my own interviews with regional stockmen of an earlier time carry information that might enhance the reader's understanding of this conflict, a problem that was fairly serious in the late-nineteenth and very early twentieth centuries. From a collection of regional folklore come the following traditional narratives on this grim subject (Attebery, *Folklore of the Lower Snake River Valley: A Regional Study*, Ph.D. Diss., University of Denver, 1961, 181–83). Hugh Kennedy (b. 1878) remembered that "Times were plenty tough. Lots of people don't know just how tough they were. Why over there in Eastern Oregon during cattle and sheep trouble there was a real war. One tough nut hired by the cattlemen, they say, killed a sheepherder and stuffed his heart in his mouth." John Lackey (b. 1880) mentioned that when Con Shea and the NG outfit were camping at McBride, Oregon, a Basque herder moved his sheep to a nearby elevation. Shea told him to leave, the Basque either did not understand or refused to follow orders, and Shea fired at the herder's feet. Shea's buckaroos then rode into the band and drove the sheep off. Lackey further recalled that in 1896 he saw five sheep camps in flames along Burnt River.

In 1898 or 1899 John Palmer (b. 1888) and his father were trailing a band of ewes to new grazing when one evening at dusk a big dark man with a sweeping handlebar mustache rode up to the camp and demanded a mutton. Mr. Palmer demurred, and the rider snorted, "Go on. We'll get one tonight anyway." As soon as the cattleman rode out of camp, Mr. Palmer sent John to warn the herders that trouble was coming.

After dark, seven men in bandannas rode up, demanded Mr. Palmer's gun, and slaughtered 300 sheep. One of the herders who was a little salty wished that "them sons of bitches would try that with me in daylight." Later, some did try the same tactics, and the boy shot one of the sheep killers and blew the hand off another.

In 1902 or 1903 a band of sheep came up through Harper and Westfall and then started down the divide between Unity and Ironside. The herders were told not to continue fooling along, that it wasn't sheep country. The herders ignored the warning, so a second warning was sent out. A third warning was delivered by fifteen or twenty cattlemen who showed up at camp with rifles.

After the first shot, which hit a frying pan, one herder fell over backward into his tent, and the slaughter began. Seven hundred sheep were killed (John Smith, b. 1885, interview, 15 August 1959. According to Martin Joyce, b. 1884, about 1,150 head of sheep were killed by cowboys west of Drewsey, Oregon, and the bones were still in evidence in 1959).

Between Crane Creek and Indian Valley in Washington County, Idaho, a band of sheep was slaughtered by cattlemen. "You bet we wore guns," said Ellis Hartley (b. 1876). "We had to." One spring, Hartley, Davis, and Leighton sheep weren't supposed to go beyond Council, Idaho. They were taken on through, however, without incident, in spite of a cattle foreman who appeared with several dire warnings.

Gale Chambers writes in the 7 August 1986 issue of the *Idaho-Stockmen Farmer* (p. 6) of a boundary line near the head of Goose Creek in southern Idaho:

> The "Dead Line," before the turn of the century, separated the sheep and cattle operations in the mountains south of Snake River. Supposedly, this was the line patrolled by Diamondfield Jack Davis, who was arrested for murdering two sheep herders John Wilson and Daniel Cummings. The trial and conviction of Diamondfield Jack Davis highlighted the old range wars between the sheepmen and cattlemen.
>
> Jack Davis was convicted on circumstantial evidence and sentenced to hang for the crime. (Supposedly, he rode some 55 miles one way, shot the two herders and rode back, as he was seen early morning and late at night on the day in question, estimated at Feb. 4, 1896.)
>
> In time, two other riders confessed to the crime, but Jack Davis was caught up in the legal mess as the argument shifted as to which legal entity had the right to hang him—the state of Idaho or the county. He spent five years in the county jail before he was turned loose—an innocent man.
>
> It was quite a bit of history, and several versions of the event can be discovered both in print and in verbal lore.

The violence happened, and like other forms of violence it was foolish, as a brief look at the causes and assumptions behind it will show.

One of the assumptions was that sheep ruin the range by eating grasses too closely and by the harmful action of thousands of small hooves destroying the tilth of the soil—tilth in the sense by which Amish farmers of Pennsylvania refer to the delicately balanced life zone of the upper topsoil. It is only partly accurate, for the presence of sheep does not change the range, but rather it is the presence of too many sheep that is harmful. This problem is the familiar one of exotic and conscienceless coyote outfits. Too many cattle and too many feral horses damage the range similarly. It should be noted here—and sheepmen have known this for years—that, in their grazing habits, sheep resemble deer more nearly than they do cattle, having a fondness for weeds, succulents, and shrubs, although both cattle and sheep flourish on white sage. Of all the herbivores, horses can probably do the severest damage since their upper and lower incisors meet and their powerful jaws make it possible for them not only to graze closely but also to pull grasses up by the roots.

I have heard cowpunchers talk about the odor of sheep, or "range maggots" as they were often called. In this matter, fecal and urine odors of all animals can become offensive if their freedom of movement is restricted. Along with this belief another element in the cowboy's value system comes into play: the superiority of the man on horseback to the man on the ground. This is such a commonplace observation that I wish not to dwell upon it but merely to remind the reader of the difference between the mounted knight and the earthbound villein. In the one instance, the noble steed could be endowed with wings, and Pegasus and Bellerophon can be paired with Bucephalus and Alexander, Traveller and Robert E. Lee, and Tony and Tom Mix. The list could be extended, of course, but the point is fairly secure that freedom, nobility, daring deeds, and a farther horizon were available to the knight, the gaucho, the Cossack, Comanche, and cowboy but not to the earthbound peasant. It is the age-old superiority—both literal and metaphoric—of the *vaquero* over the *bracero*. Naturally, cowboys looked down on sheepherders and, conversely, sheepherders looked up to cowboys.

Beliefs such as these are best tempered by example, not precept, and that example is provided by stockmen like Phil Soulen, who is both cattleman and sheepman. No hostility inheres in the nature of these enterprises. In the dry American West, both depend upon sound range management, common sense, and sometimes generous—sometimes stingy—Mother Nature. One dry year with sparse forage can force the operator to employ the survival strategies validated by tradition from time immemorial—selling more ewes than usual, selling ewe lambs otherwise intended as replacements for mature ewes, selling more young wether lambs under 125 pounds, arranging for more irrigated pasture, buying and feeding (or attempting to buy) more hay, buying and feeding supplements. Market and climate set in motion these traditional patterns by which the stockman attempts to survive.

Regional newspapers devote features from time to time to the problems and prospects of various segments of a state's or a region's agrarian pursuits. There is a certain irony in all this because such periodicals are usually published in a highly urbanized setting where a recurring problem is the availability of inexpensive groceries. Indeed, to many metropolites, the more remote rural corners of their state or region exist as either *terra incognita* or *deserta*, for it is increasingly unlikely that there is a generational tie to the land. The flight to the city occurred so long ago that any given family's connection to the land is tenuous.

At least, that is a common perception, and a case in point is the fracture between inhabitants of the wet part of Oregon, who are numerous, and those of the dry part, of whom there are few, although theirs is by far the larger land mass. This fracture is apparent in the names applied to each group: Webfoots vs. Bunchgrassers (Suzi Jones, *Webfoots and Bunchgrassers: Folk Art of the Oregon Country* [N.P.: Oregon Arts Commission, 1980]) or Meier and Frankers vs. Eastern Oregonians, as I have heard Dick Miller and others refer to the difference. Meier and Frankers, named after one of Portland, Oregon's, pioneer and quality stores, are hunters who drive east to harvest pheasants or deer, elk, and antelope in the dry country east of the Cascades. And as the metropolitan hunters pursue wild game in dry eastern Oregon, journalists will occasionally turn to this same remote country for the features they can harvest.

I must affirm here that there is nothing inherently wrong or evil in this phenomenon, which the eighteenth century might have called the "economy of nature." If there is a *sense* of wrong, it lies in the perception of exploitation, that one part of the country exists for the support of another. Indeed, this sense of exploitation has been used to describe the relationship of the entire American West to the cannibalistic East, the assumed cultural, communication, fiscal, and political power center of the nation. To enter that arena would take this study of tradition and sheep raising far beyond its reasonable limits. But the matter cannot be entirely overlooked, inasmuch as parallel problems may exist within the regions of the nation.

For instance, "The Northwest" section of *The Oregonian* for 13 March 1988 featured an essay entitled "The Comeback in Cattle Country" by David Braly, about how contemporary cattle ranchers in eastern Oregon have learned to survive. The revelation in the essay is that suddenly technology, the means to survival, has been capitalized upon in embryo transfers, in the use of computers, and in taking advantage of the market which now calls for leaner beef. An item in a local newspaper of 23 March 1988 discusses embryo transfer technology for sheep in research conducted by the University of Idaho. White Panama ewes gave birth to purebred black Suffolk lambs, which will mature as white sheep with black faces (*Idaho Press Tribune*). But it is an old story line, even though the particularities of plot vary. The alert reader will recognize the familiar theme and variations or dynamic variability, to use synonyms.

I would suggest that livestock growers have consistently employed any or all means capable not of guaranteeing survival—for there are no guarantees on this planet—but of making survival likely. This is not the place to document technology-inspired variations in the livestock industry, but even the casual observer will note, with respect to cattle raising, that trail drives to market yielded to improved technologies of transportation as railroads became available. The same observer will recall further that poor beef varieties—longhorned cattle—were replaced by better varieties as the technologies of raising purebred stock proved profitable. The list goes on, will continue to go on, and it is not likely that any one technological advance will signal that growers of livestock have achieved guarantees of success. The prospect of genetic engineering, of recombinant gene-splicing, as noted earlier, is an entirely different matter.

So the question recurs as to whether the basic reality of raising livestock, their management, their access to pasture, their marketing, has changed. It appears to me the question deserves a negative answer. Even with artificial insemination, embryo transfers, and computerized accounting, and market quotations across the livestock industry generally, the basic reality is constant. Technologies, instrumentalities of the operations, change, but those are externalities.

Harry Smith of the "CBS This Morning" show recently (13 May 1991) showed a taped interview with a western ranch couple about pressures on public lands. Seeking to balance the perspective, he allowed time for ripostes from a speaker for an opposing point of view who said, in effect, "They [ranchers] think because they've lived here forever they have a right to abuse it [the rangelands]." A peculiar function of time and technology has brought this matter of traditional access to public lands by ranchers into question.

Few doubt that enactment of the Taylor grazing legislation was essential to the commencement of rehabilitation of badly mistreated public grazing lands. It does little good, however, to malign contemporary ranchers for this abuse. No amount of historical revisionism will eliminate the sins of slavery, of destruction or debasement of Native Americans, of the defoliation of forests, and the gobbling of oil and mineral riches, which my grade-school geography books described as unlimited. In 1934 the word recycling had not been invented.

From 1934 (while ranchers were paying the fees required for grazing cattle and sheep, helping develop and maintain springs in a land of precious little rainfall, and fighting range fires in cooperation with the Bureau of Land Management) until about 1961 or 1962, there was relatively little nonlivestock pressure on the land. The multiple-use concept of management made public lands available to photographers, rock hounds, and others. Then, by the late 1960s began a crescendo of use that continues to increase. An improving national economy; the greater availability of reliable four-wheel-drive vehicles; improved backpacks; better hiking gear; freeze-dried, affordable, palatable food; and lighter tents and sleeping bags—all these technological advances (at an unspecified cost in energy) made the inhospitable semiarid BLM lands available to a public, an urban public, increasingly in need of the emotional restorative to be gained from lonely places under a wide sky.

Perhaps the point bears repeating: From 1934 until the technological advances evident by the 1960s, the BLM lands were singularly unattractive as recreation sites and were thus largely ignored by urban folk. But all this while the ranchers endured, living far from anywhere, persistently doing the best they could, making the best of what they had, and generally following the guidance of the BLM. Now that the technologies of outdoor living have enabled urban dwellers to control the dry and spacious West, they venture into regions where but for technology they would probably have never set foot. It is the difference between home and vacationland.

One further observation needs to be made. The "CBS This Morning" guest who spoke against those who "think because they've lived here forever they have a right to abuse it" was right in demanding no more abuse. But such a perspective is clouded when the view sees the western landscape in relation to lands east of the one hundredth meridian. It is for this reason that the argument about livestock production in the West as compared to production in the rest of the United States is unsound. That argument usually takes the following form: why should we be concerned about the western livestock industry when it produces only about 4 (or 6 or 10, whatever the accurate figure is) percent of the beef produced in this country?

Such an argument would appear to be slanted if not downright unfair. It fails to recognize that most of the population of the

United States is east of that crucial meridian. Should that fact cause the continent to tip always east? The West already has fewer congressmen than the East; should the number be further reduced? The numerically inferior western livestock industry helps support the numerically inferior population of the West. If only numbers matter, the sparsely populated West exists only to be cannibalized by the East. Never mind that a way of life has been developed by a few people adjusting to an environment that is indifferent to a human presence, a way of life that produces food and fiber for the public and profound job satisfaction for the few who have perhaps earned the right to maintain this traditional connection with the land. Nowhere does this right presuppose the right to abuse, however, for with abuse the renewability of grass and water is severely compromised.

More years ago than I care to dwell upon, I wrote an essay which I called "The Fiddle Tune: An American Artifact," and it eventually appeared in a collection of folkloric materials edited by Jan Brunvand entitled *Readings in American Folklore* (New York: W. W. Norton, 1979, 324–33). Among other ideas in the essay is the suggestion that a great many of our artifacts—broadly defined—reveal that Americans have been fascinated "by activity, by movement, by process." Respecting the fiddle tune, the essay calls attention to the fact that there is no internal necessity or structural requirement that the tune must stop at an aesthetically determined stopping point. It stops, therefore, when the fiddler wearies or when convention decides the dance set should end. Otherwise the tune would continue indefinitely. These endings are often signalized by a highly artificial coda, "shave and a haircut, six bits."

Raising sheep in the tradition of transhumance has not ended, and my personal examination of it ought also to be ongoing so that the tension of variation within continuity might be described. Alexander Campbell McGregor was able to write a history of his forebears' sheep raising because that family quit raising sheep. The Soulen enterprise continues; its activities continue to be described, for the reality of transhumance as practiced by this family operation will be characterized by both theme (constancy) and variation.

Great-grandchildren of Harry and Beulah Soulen are more than casual observers at shearing and shipping times.

But the personal examination by the essayist must stop, after extending over more than three years. The stopping is not a conclusion; it is a halt. Instead of a coda as satisfying as "shave and a haircut, six bits," I offer only a wish and a final paragraph. The wish is that sheep may safely continue to graze and pasture.

Annually, to the delight of journalists faced with column inches of white space that must be filled and to the pleasure of sheepmen, who like to watch their ewes graze safely and their lambs suck in contentment, sheep move to new pastures as life reappears in vegetation and in animal progeny. It is all very much like a folk song, and in a sense like life itself, a recurring pattern of theme and variation.

Appendix A

(The following "sidebar" by David Proctor [the *Idaho Statesman*, Business Section, 20 September 1987, "Shear Trivia about Sheep"] provides a useful summary of pertinent information about the sheep business.)

The sheep industry goes a long, long way ba-a-a-a-ck:

Sheep were domesticated in western Asia about 10,000 B.C., and there is evidence of domesticated sheep in prehistoric Europe.

During the Middle Ages, England and Spain were at the head of the flock. Sheep came to the New World via South America in the 16th and 17th century.

They came to western Idaho from California via Oregon and to eastern Idaho from Colorado and Wyoming about the time of the Civil War. By 1867, the breeding sheep population in Idaho hit 14,000; by 1890, it jumped to 614,000. Sheep numbers peaked in 1918 at 2.65 million, nearly six times the human population. Nationally, the sheep population peaked at 56.7 million head in 1942. In 1985 it hit 9.9 million.

Four breeds were developed in Idaho. The U.S. Sheep Experimental Station at Dubois created the Columbia in 1917, the Targhee in the 1930s and the Polypay in 1969. The Laidlaw family developed the Panama breed in the 1950s at their ranch in Muldoon Canyon near Carey.

The Taylor Grazing Act, passed by Congress in 1934, divided the federal range among the segments of the livestock industry to correct over-grazing problems. Sheep numbers began to drop.

After World War II, the use of synthetics and a shortage of sheepherders also contributed to declining numbers. The numbers of Basque sheepherders were reduced by better economic conditions in Spain and tighter U.S. immigration laws.

Lamb consumption also dropped from 5 pounds per person in 1963 to less than 2 pounds in 1983, though it has increased slightly since then. In comparison, Americans each eat an average of 63 pounds of beef annually.

In 1961, there were 1 million head of breeding stock in Idaho and 670,000 people. Human numbers finally exceeded sheep numbers in the 1970 census with 687,000 sheep and 700,000 people.

In 1980, there were 456,000 sheep and 713,000 people. In 1987, there are 299,000 head of breeding stock and about 1 million people.

In 1931 it was estimated the sheep industry produced an annual wool and lamb income of $30 million—enough money to spread a double line of dollar bills across the widest part of the state and still have some left over. It was the third largest industry in the state.

In 1987, the income estimate is again $30 million, though the modern dollars will not stretch as far. The industry's total impact on the state—for trucking, feed, wages and so on—is $120 million.

APPENDIX B

(The following extended essay comes to me from Stratford, New Zealand, where Mrs. Jeanne [Poulton] Coles, whose husband, W. R. Coles, raises sheep, has conducted her own study of the theme and variations of sheep raising. The statistical figures are from the 1987/88 season. Mrs. Coles is an alumna of The College of Idaho.)

All lambs born in the New Zealand spring—August and September—have a future of some sort. Because the climate is relatively temperate, from about 15 degrees upwards in winter, all lambs are born in open pastures. Cold isn't as dangerous as wet with cold, and it often seems that more lambs arrive on the coldest, wettest nights. Lambing season once meant plodding round and round the "paddocks" assisting ewes in distress or "mothering" orphaned lambs with foster ewes if their own ewes have died or wandered off with the stronger twin.

Sometimes a farmer can simply enjoy good spring days. Some farmers have found that, by strict culling of ewes with a history of lambing trouble or general poor health, the plodding and plodding of paddocks in spring isn't so necessary as once thought. In fact, often the strange vertical figure in oilskin raincoat and leggings, making peculiar noises, or putt-putting on farm motor bikes, causes more harm than good. It can cause ewes to separate from weaker lambs.

New Zealand lambing percentages average about 95%. Interestingly, the South Island has nearly 100%; whereas, the North Island has about 90%. There's a theory this is due to the longer daylight hours nearer the South Pole at conception time. "Mothering up" is done in several ways. Some farmers tie the ewe to a nearby fence with that most essential commodity, baler twine, so she can't get away from the hungry lamb. Some use a chemical cream rubbed on lamb's coat and then on ewe's nose to confuse the scent; some put lamb and ewe together in a small pen. Most farmers' wives feed foster milk to a few lambs until a nurse-ewe is available.

If it survives lambing time, the young lamb's next decisive hurdle is docking time, carried out at any time from a week to eight weeks after birth. All lambs lose their tails; methods vary. Some farmers use a sharp knife, or a hot searing iron, but the most popular method is to attach a constricting rubber ring to the root of the tail, cutting off the blood supply and causing the tail eventually to drop. Farmers count the number of tails or rubber rings used, to tally lambing percentages.

On stud (registered) breeding farms, certain lambs are left to grow to ramhood. But the vast majority of the ram lambs born get a second castrating rubber ring, which destines them for the fat lamb market sometime within the first five months of life. Thirty-five million fat lambs are anticipated in this coming farming year. Ewe lambs get a longer life. All lambs are ear-marked for sex and for ownership. Ewe lambs get another ear-mark later in life to indicate their age.

Docking is rather a sad time. The lambs that frolicked amongst the daffodils growing in the paddocks don't frolic for quite a few days. They're recovering from their injuries and from routine injections they get for blood poisoning or pulpy kidney, diseases which can suddenly kill the healthiest of lambs. Or they may have a vaccination for scabby mouth, a virus from the ground that causes the nose and mouth skin to peel off.

Every sheep farmer (and quite often his wife, whose job it is to catch and hold the lambs) hopes for dry weather toward the end of September and October to get the docking pens set up and another mob done. Two hundred ewes and lambs in one day is enough for a husband and wife team. Those with more help can do five hundred or more a day.

Tricks learned from years of experience help make the job easier. Long lengths of gunny sacking, called "scrim" in New

Zealand, help drive the stubborn ewes and lambs into a pen in the corner of a paddock fence. It's best to move the sheep slowly and quietly until a crucial point, where the more noise, the better. A gap of wire netting that the ewes can see through at the end of the pen helps draw them forward, as they don't want to approach what they can't see through. The lambs are separated from the ewes into a small catching pen about six feet square.

After docking, farmers concentrate on shearing dry sheep, or perhaps they have a lull in work. But within two months, the summer vigil of drenching begins, as New Zealand's wet and warm climate is particularly salubrious for worms. In wet areas the lambs need a mouthful of worm drench monthly.

In the last few years another pest called "fly strike" has come, perhaps the most unpleasant in the whole business. A new, small green fly has arrived in New Zealand from somewhere and blows its eggs on a sheep, anywhere, any age. The eggs hatch and the maggots begin to eat. Farmers spend many hours wandering round their sheep looking for the signs of maggots at work under the wool. The wool has to be shorn off the infected area and strong medicine applied. It is disgusting and sad, and will continue to get worse, for there's only limited control for such things as flies coming in from the sky.

Woe to the lambs destined for the fat lamb market. The best price per pound is for lambs at from 33 to 42 pounds dressed weight. That's about 66 to 84 pounds live weight. The few sold to the United States are the larger lambs. Last season the prices for these weights were about $14 each, New Zealand dollars, which is about $9 U.S. (3–15–88). This price includes about $4.90 U.S. for the skin. Even after weaning off mother ewe's milk at about three months, lambs will reach those weights within about four months.

Freezing works hire fat lamb buyers to travel round farms marking the lambs ready for killing. A gentle squeeze on lamb's rump will indicate his readiness. Unless the employees at the works are on strike (I give them credit for not striking so much the last two seasons), the lambs will be on their way within hours. Double-decker trucks hold up to 400 lambs; triple-decker trucks 660. The left-over lambs are drenched and put back to pasture for another two to four weeks.

Somewhere between the ages of nine and fifteen months (nobody really agrees as to when), lambs are classed as "hoggets." Nearly all the mutton eaten in New Zealand is from this age animal. These hoggets will have been shorn two or three times by their fifteenth month. They do better without their wool. At eighteen months, sheep get their second teeth and are then called "two-tooths."

Though older sheep do not need the constant attention given lambs, most farmers still give ewes worm drench annually. The old-time dipping was done by plunging the sheep into a trench filled with water and chemicals, with someone standing over to poke the sheeps' heads under as they swam through. There are alternative methods, such as running the sheep through a series of sprays or putting them in a small, round pen with sprays under, and rotating above. The advantage of the latter method is that run-off spray material can be strained and used again, a wise move, as the cost of spray chemicals and drenches is very high. Nowadays, pour-on lice chemicals are becoming popular.

There are about seventy million sheep in New Zealand, but the human population is just over three million. An average family sheep farm would have from 1,200 to 1,500 ewes. Larger sheep farms with many, many more sheep are called "Stations." Most sheep farmers' income is generally more from wool than from sheep meat, depending on the prices. Consequently, every sheep farm must have a wool shed for shearing or share with a close neighbor. Most sheep are shorn once or twice a year. Full-year fleeces are about 8 inches long; a whole fleece weighs about 12 pounds. Some shear three times in two years, when the fleeces are about 5 inches long. The price for wool this last season was about $1.40 U.S. per pound.

Shearers now get 72 N.Z. cents (47 cents U.S.) per ewe or lamb, and better shearers can do about 300 or more per day.

It's good money, but it's also very hard work. There aren't any lazy shearers. They begin at 5:30 A.M. and go till breakfast at 7. Again from 8 A.M. until morning tea at 10:15—called "smoko" here—sandwiches, bacon and egg pies, cakes, etc. They work again from 10:45 until 12 for lunch; from 1 until 2:45 when they break for "smoko"; from 3:15 until 5:30: a nine hour day.

The heavy overnight dew in New Zealand means sheep can't be left out the night before shearing, so sheds must have provision to cover enough sheep to keep the shearers going until noon the next day. Farmers always hope shearing days will be warm enough to dry the dew off the other sheep by then.

Besides hiring shearers, a farmer must supply a "fleeco"—usually two people per three shearers—to keep the wool picked up from the shearing board. A "board" really is a floor, though often raised a couple feet in newer sheds. Then the fleeco must pack the wool into a bale in a wool press and tramp it down very tight. The wool is hydraulicly pressed and sewn into 350 to 400 pound bales about four feet high by two feet square. The fleeco must also keep the shearers' catching pens full of unshorn sheep. A shearer, the "king of the shed," doesn't like any of his working time being wasted looking for his next customer.

"Dagging" is another not-so-pleasant aspect of sheep farming, for "dags" are accumulations on the "behinds" of sheep. This faecal matter ensnared in the wool is especially prevalent in a season when the grass has grown fast. Dags have to be shorn off whenever they collect, especially in the "fly strike" season. Strangely, there are colloquialisms that have come from these not-so-pleasant things. The expression "rattle your dags," means "hurry up" for when a dag-laden sheep hurries, its hard, dried dags, "rattle." To call someone a "dag" means he's a joker, a particularly jolly sort of fellow.

Grazing ewes are useful for cleaning "paddocks" (fields). They eat weeds that cattle won't eat; if pushed, they crop the grass shorter than cattle, which helps prevent the grass grub beetles and worms from getting into the grass pastures and eating roots. Ewes climb steeper hills; they push through fallen "scrub" (native bush) to open it to decay and the sun, so grass can grow underneath. Just before time for rams to be let out in March, the ewes are put on good pasture to bring their condition up. This makes for a better lambing percentage. They may need to be crutched before this time, which means yet another time "over the board" in the shearing shed. Unless they've been shorn before lambing they'll need crutching then, too, to clear the wool from the behind and underneath area so the lamb can better find his milk. Sheep farmers need good backs.

For those who shear in winter or before lambing, storms can be a worry. It's important that the ewes have time after shearing to get a good feed of grass. They can tolerate cold, but cold and wet are very dangerous. In one or two days they're reasonably adjusted to their loss of wool. Pre-lamb shearing has its advantages, even though many think it's cruel. The ewes will seek shelter for lambing; they will not need to be handled again before lambing for the otherwise necessary crutching. This pre-lamb shearing also avoids the "break," the weak place in the wool fibre caused by lambing and winter ills. Shorn sheep don't so easily get "cast"—on their side, unable to get up, in which position they die very quickly.

So—a year's events in the life of a sheep. If it be a ewe, she may live five to seven of these yearly cycles, to be eaten in the end, quite likely by Japanese. But if it be a ram, he might be eaten within his first cycle by an Englishman, an Arab, or an American. Or if he's born into an "elite" flock, and has spent his life doing his duties as a sire, he'll most likely end up some six or seven years later as dog tucker.

Appendix C

(These snippets from the Emmett *Index* and the *Examiner* reflect the importance of sheep raising in an earlier time and the availability of that activity as news. The same kinds of stories can be found in just about every newspaper in the region.)

Emmett *Index*
17 February 1907
James Little will this year try early lambing as an experiment. He has built long rows of sheds on ground protected from the winds by trees and is prepared to keep his ewes and their babies as snug as a bug in a rug. He has close to 2500 ewes. The lambs will be fattened and shipped east for the spring market.

Emmett *Index*
9 April 1908
Over two tons of powder and 100,000 caps have been purchased this spring from one hardware store in this city—D. A. Hawkins—by the owners of sheep to protect their flocks from coyotes and other wild animals during the lambing season. Purchases of explosives have also been made from other dealers, which gives some idea of the expenses incurred at this season of the year to intimidate those animals which have an insatiable appetite for a lamb on the half shell.

Emmett *Index*
13 July 1911
Trainload of fat sheep and lambs headed for Omaha and Chicago. 47 doubledeck cars—12,000 animals—joint shipment of Andy, Walter and Jack Little—all prime stuff. Several

men including Littles accompanied sheep to Omaha and Chicago. Van Duesen Bros. to make shipment this mo. Sheep market strong. (Gives prices in article.)

Emmett *Index*
20 June 1912
Forest Service to make test of acreage necessary for sheep. Andy Little has placed at department's disposal 1,250 head of sheep.

Emmett *Index*
11 April 1912
Andy Little lost 130 head of sheep on Four Mi. Creek from eating the roots of wild parsnips which it is claimed are rank poison.

Emmett *Index*
6 July 1911
Mr. Little one of the most extensive sheepraisers and wool producers in this state and a man of large business interests—his wool clip this spring sold for 1/2 million dollars.

Emmett *Index*
11 April 1912
Andy Little's team came in from Big Willow Creek and bought 3,800 pounds of sugar, flour, salt and oats from Mr. Babcock. He has a good trade with the sheepman.

Emmett *Index*
16 May 1912
Sheep shearing is on in earnest in this section and the Fruit Assoc. warehouse is filling up with the bags of wool that are coming in. Andy Little's crew of Mexicans arrived Saturday morning and at once started work. The clip this year is lighter than usual.

Emmett *Index*
10 October 1912
Through the carelessness of a herder, Alex Cruickshank of Struan Hill Ranch, e. of Montour, on Tuesday suffered the loss of 570 head of ewes, which went over a hill into a gulch and piled one on top of the other. The sheep were fine animals and the loss is a severe one to Mr. Cruickshank.

Emmett *Examiner*
5 January 1911
Andy Little, John Van Deusen and A. Cruickshank, all prominent sheepmen of this section, are representing Emmett, at the meeting in Portland, this week, of the National Wool Grower's Association.

Emmett *Examiner*
28 April 1910
Andy Little, Emmett's big sheepman, will commence shearing his immense flocks of sheep on Willow Creek, about 12 miles southeast of town, tomorrow.

Emmett *Examiner*
5 May 1910
Jack Bruce, the big sheepman of Upper Squaw Creek was an Emmett visitor last Friday. He reports flocks to be in fine condition, and ready to be de-hirsuted.

Emmett *Examiner*
5 May 1910
McConnel brothers, who have charge of the Anderson Creek ranch, about 7 miles east of Emmett, state that their two bands of sheep will show a lambing increase of 100% this spring.

Notice to Sheepmen
Emmett *Index*
30 April 1914
Notice is hereby given that anyone violating the Two Mile Limit Sheep Law of the State of Idaho, along the premises of the undersigned, will be prosecuted to the full extent of the law. (Signed): George Stephens, Robert Cornett, Fred Mohr, Lafayette McFadden, William Walker, Chas. Walker. Dated March 25, 1914.

Emmett *Index*
22 July 1915
Andy, Sam and Walter Little are this week shipping their entire wool clip to the Koshland Company of Boston. The consignment consists of approximately half a million pounds and will fill 14 or 15 cars.

Emmett *Index*
19 August 1915
Andy Little is now justly entitled to the name of the Sheep King of Idaho. This week he closed a deal, whereby he bought the entire holdings of Scott Anderson of Boise. There were 25,000 sheep in the bunch, besides several sheep ranches in the Boise Basin country. It is stated upon good authority that during the present month Mr. Little will ship to the market about 70 cars of lambs, totaling 21,000 lambs. There are 300 lambs to the car.

APPENDIX D

(Folklorist-architectural historian Jennifer Eastman Attebery photographed the Soulen home at the south end of Payette Lake, McCall, Idaho. The following information developed, in part, from her interview with Beulah [Mrs. Harry] Soulen.)

The Soulen's upper pasture operation is based in McCall out of their house on the lake. Built in 1934, according to Beulah Soulen, for her and Harry Soulen, the house is Rustic Style, built of logs that were hand hewn by local labor and laid in even tiers with the log ends projecting beyond the joint. No notching. The Rustic Style is apparent in the exposed roof beams, exposed logs on the interior, and multipaned windows. On the ends of the stair and roof beams the architect carved ram's horns. According to Beulah, the architect, who also supervised and participated in construction, was named Semro and was from California. She asked him to do their house when he was in McCall employed in the design of the Sylvan Beach buildings, now owned by J. R. Simplot. Except for recent kitchen remodeling and reglazing of a set of multipaned windows with one large picture window on the rear wall, the house is architecturally intact, and probably eligible for the National Register.

Harry and Beulah Soulen got into the ranching business by buying up an outfit in the 1920s, and they began using the upper pasture in the late 1920s, living in a small cabin on the lake that they built and in other rental cabins and houses in McCall. When they built the present building on the lake the position of the house close-in to town was partly determined by the fact that telephone lines only extended as far as the house's location. Telephone service was important in maintaining contact with shippers, markets, and suppliers. It was unusual to build on the lake at that time. Houses were set back and facing the streets or perimeter roads.

The home of Harry and Beulah Soulen in McCall, Idaho.

Harry and Beulah Soulen, parents of son and present owner Phil Soulen, built this home in 1934. It is the summer headquarters of the extensive Soulen Livestock Company.

Appendix E

NARRATOR: Stewart Cruickshank
DATE: 2 October 1986
INTERVIEWER: Madeline Buckendorf
LOCATION: Parma, Idaho
SUBJECT: Montour

This is part of an interview of Stewart Cruickshank, conducted at his home in Parma, Idaho, by Madeline Buckendorf on 2 October 1986. This interview was part of a cooperative project between the Bureau of Reclamation and the Idaho State Historical Society.

MB: You told me a little about your grandfather coming to this country?

SC: He came over here in the late eighties, into Canada from Scotland. He was apprenticing as a butcher in Glasgow, and he was not the oldest son, so there was... in Scotland, the oldest son got everything, you know, and he wasn't the oldest son. So he was apprenticing as a butcher in Glasgow, and he took some meat down on the dock, and they asked him if he wanted to come take care of livestock on a livestock boat coming into Canada, and he left the cart right on the dock and came to Canada. Well, his family never heard from him for several years, but he didn't... he hated the royal family of England, and they were too much in Canada, too much royalty in Canada, for him. So he migrated into Nebraska, and homesteaded there, or at least acquired property there, and this was probably in the late—or about the middle nineties, 1890s—that he went into Nebraska. But it was... my father was born there in '97, in Nebraska. Or all his family, all his brothers and sisters were born in Nebraska, and they migrated to Emmett, then in 1904, in an immigrant train. They got into Emmett, he told me, the seventeenth of March in 1904. And my dad was seven years old when they came to Emmett.

MB: Okay. Then how did your grandfather get a little interested in sheep?

SC: Oh, they were just livestock people, and the area at that time was basically sheep area, the Emmett-Payette River area. There wasn't too many cattle in the area at that time, and the Littles had a big sheep outfit there, were getting into the sheep outfit, and the Van Duesens were old-time sheepmen in the area. Skillern was another old-time sheepman. My dad used to help them load sheep. At one time, Emmett was a railhead for all the sheep in that country, when they shipped their lambs in the fall of the year, that they would come into Emmett and ship their lambs, and he had a pet sheep that he led them on the cars with.

MB: When your dad was up on his first place he homesteaded, besides the sheep, did he raise anything?

SC: No, just sheep. No, just range, just sheep, that's all. That's all we had was sheep. Then after we moved down to the other place—and I started to school fall of 1925—then along about 1927, '28, along in there we started raising turkeys. And that lasted up until, oh, '37, '38. Probably a ten-year period. We raised turkeys during the Depression to stay in business. (Laughter)

MB: Yeah, you said something about that paid for the sheep?

SC: Well, it bought a lot of land. My dad was starting to accumulating buying land; it bought a lot of land, and it just provided—the sheep was on their own—and those turkeys provided the money for us to live on, plus accumulated enough money to send my sister and I to college.

MB: That amazes me that it was able to do that during...

SC: The Depression.

MB: During the Depression.

SC: But you want to remember that my sister and I went to eight years of college on $4,000. (Laughter)

MB: That's true.

SC: We had $2,000 apiece, and we went—we both of us graduated from the University of Idaho. We had $500 a year for board and room and books.

MB: Those days are long gone. Going back a little bit into the sheep, and then we'll talk about the turkeys a little bit later, when your father started really getting into the sheep, this was in the twenties then?

SC: Well, he really was in the sheep on his own by 1916, and then he... after him and my mother was married in 1919 he had his own sheep then, and... let's see. In 1919 was when he bought in with his father. He'd sold his sheep there and—the ones he owned himself, a band of ewes—in 1918, because he thought he was going to have to... he'd be drafted into the army and he sold his sheep. And when he come back from that, he bought in with his dad with the money he got for that band of sheep.

MB: Would you explain to me a little bit about the difference between range lambing and shed lambing?

SC: Well, range lambing is—they usually do it—in this area there aren't any range lambers left. They would lamb those sheep in April, on the range. Just spread them out and lamb them, out there. Well, shed lambing, they usually did it in January and February under... when they were on the feed yard, and they had lambing sheds, and they saved a lot more lambs. The percentages went up quite a little bit from where you'd have a 75 to 80 percent lambing on the range; you could get 100 and 20, 25 percent in the lambing shed.

MB: Do you know why that change occurred?

SC: Oh, it was just a change of management. It was to take advantage of the market, because traditionally the lamb market is higher in May and June, and by lambing in January and February you could get that earlier market. [Phone rings in background.] What were we talking about?

MB: About... I thought I might ask, once you went to shed lambing, did that require more help, or any other changes?

SC: That required more help, in the wintertime. You had to have more help. It took a little more help than range lambing. It took quite a crew range lambing, 'cause you had to—you know—you had to separate those ewes that are lambed from the ones that hadn't lambed, and try to catch the ones that weren't suckling, and those kind of things. But it took more help in lambing sheds, because you just had to—it was more intensified. You intensified your lambing.

MB: Where would this help come from?

SC: Just local help. Usually, oh, all these small farms, in those days, on small farms down here in the valley in the wintertime, you know, there's... they all wanted to work. They didn't have any other work in those days, because the cash crop in this valley at that time was hay, and they sold it to the livestock people to winter on. That was the reason that the livestock came down along the Snake River, because of the feed supply in the wintertime.

....

You see, your sheep numbers peaked out in 1941, or about 51 or 52 million, and by 1960, we were down to 15 million.

MB: I also thought the twenties was kind of a rough time for the sheepmen too, when...

SC: That's right, the early twenties, along '21 and '22, and when a lot of sheepmen went broke. There was a lot of big outfits. Like the old Clinton outfit was one of them in this country, and they run a lot of Scotchmen with sheep, especially in the Jordan Valley area and Owyhee County, and over in the eastern Oregon. And they went broke. And that's when the Basques—Bascos—first started getting into the sheep business, because the banks turned it over to them herders, 'cause they didn't want them, and they'd just turn them over to those herders, and that's what put the Bascos in the sheep business. That was in the twenties.

MB: When you talk about bands, when you had a band of sheep, you had several bands moving through?

SC: That's right.

MB: What was the size of those bands?

SC: Oh, in a spring band you had 1,000 to 1,200 ewes and their lambs, which would make them around 2,400 or 2,500. Twenty-five hundred to a band. And in the fall, bands was around 1,000, 1,800 to 2,000 ewes in the band.

MB: And so you'd have those separate bands moving through?

SC: Mn-hmmm. Usually you have two men with a band. Some outfits run three men to two bands, one camptender and two herders to two bands, but we always had two people with every band, a camptender and a herder, and a pack string.

MB: What... how about dogs with that? Did you have dogs?

SC: Oh, you always had three or four or five dogs to the band, you know, two at the minimum.

MB: With the dogs, did you have a special kind, or did you...?

SC: Oh, most of them were some kind of shepherd dog, collie dogs, English shepherd or border collies or sable collies. Those kind of dogs—shepherds.

MB: Where would those come from? Were they kind of in the family, or...?

SC: Yeah, everybody had a line of dogs, a string of dogs, you know, they'd be some guys preferred English shepherd, some of them borders, and they were all mixed up, those old sheep dogs was. They was all kinds, and a lot of Lassies. But our dogs, mostly, were English shepherd, and woolly-faced ones. And then in later years, that string kind of run out, and then we was into border collies. Natural-born livestock dogs.

MB: What kind of problems would you have along the trail, when you were out trailing them?

SC: Oh, finding feed and water was the big problem. You know, there was so many trail sheep in those days, back and up until way up into the fifties. You had designated trails, and they'd just be dust beds, when there was a lot of sheep in the country. And you just had to find feed and water when you was on the trail, was a big problem. And keeping ahead of the guy behind you, and not trying to run over the guy in front of you. (Chuckles)

MB: Keeping things spaced out.

SC: Yeah, stretched out.

MB: How would you do that? Was there any planned order, or just...?

SC: No, you just took your turn. I mean if you came to a bottleneck in the trail, why you just took your turn going down that, you know. There was a lot of help between one another. Usually the camptender rode in the lead, so that they wouldn't mix, and these sort of things.

MB: Now, have the shearing crews... how is shearing handled? Has that changed over the years?

SC: Oh, it's changed a lot over the years. As the numbers went down, the crews got less, and it got to where they couldn't make a living. Those old shearing crews would start in California, you know, and shear clear into Canada and make a living at it. Well, it got to where they couldn't make a living at it, and it's just a kind of a helter-skelter deal now. You get a lot of Mexican crews, come out of Mexico. There's even some New Zealand crews in the country now, if the government don't catch up with them and deport them. But there isn't... there is one thing that's really developed over the years. There isn't any real good sheepshearers left in the United States. There's a few, you know, there's always a few, but there just isn't enough sheep left anymore for anybody to make a living at it, so you just use part-time people, or these old sheepshearers, old Social Security guys. They still come out and shear about one-third or a fourth of what they used to shear. Take them all day to shear 50, where they used to get 200 a day. And this is one thing in the sheep business today, that's a problem, is finding sheepshearers. There's no young ones. Except there's young Mexicans and young New Zealanders.

MB: I'd heard something about Peruvian—maybe those were herders.

SC: Those were herders. That's what they're using on this immigration program, to get herders, is they're using Peruvians.

MB: Early on, who were the herders?

SC: Most of the herders that we had in this country came out of... were Spanish Bascos, and people out of the mountain—out of the Appalachian plateau country, Kentucky and Tennessee and West Virginia and the Carolinas. There was a lot of those people out here in the... prior to that there was a lot of Irish and Scotch sheepherders. And then there was all kinds. I never saw a nationality that wasn't herding sheep. I've seen Portuguese, and Italians, and Germans and Russians and Norwegians. There was all nationalities. I can't think of a nationality that didn't herd them, at some time or other. And there was a lot of... at one time during the Depression there, that a lot of well-educated people [were] herding sheep. Fact is, we had two schoolteachers herding sheep. And one of them was a graduate of the University of Montana that was teaching school, and he could make more money herding sheep than he could teaching school. And there was another one that was a graduate of the University of Kansas that could make more money herding sheep. It was a good job in those days, when sheep was king. They got... now I can't remember. There was some wages as low as $15 a month, and board and room, you know, but I can't remember my dad ever paying less than $30—$30 a month.

MB: And board and room.

SC: Board and room, yeah.

MB: Where would they stay?

SC: Oh, in sheep camp. You had your tent and your camp right with the sheep. Your home was there, everything was there. The kitchen, your bed, the tent. You moved it with a mule, every day, every other day or something. You know, you was constantly on the move. You see, we migrated. We were what they called migratory livestock, in the taxing system, our state taxing system, you know. And we moved from the Snake River clear into almost Idaho County. It was a hundred—about 150 miles—from our lambing sheds to the head ends of our ranges in the Deadwood ranges, and we made that trip twice a year, that migration, up and back. Walked them all the way.

MB: Last time when I talked to you—this is jumping back a little bit—you told me about in the twenties, when a lot of groups went out of business, about a group called—or people called "coyote outfits?" Now what were those again?

SC: Well, a coyote outfit was a—what we called a coyote outfit—they didn't own any land. They had no base, really. They just run those sheep on the public domain, and when they... that was before the Taylor Grazing. The Taylor Grazing put all those outfits out of business, or forced them to buy land, in order to get a permit under Taylor Grazing. And they just wandered around wherever they could go, you know, wherever there was open land. Then when they'd run out of that, they would go into Long Valley or some area like that, to summer, buy pasture in the summertime. And then when that was over with, they'd come back down to the open ranges, out of the boundaries of the Forest Service. Now that lasted from the time—from the twenties until...oh, in '33, probably to '36, '37, when the Taylor Grazing and all was in place. I think '33 was when the Taylor Grazing Act was passed, and it was probably '36, '37 before all those outfits were gone. There was a lot of them in Nevada. They went clear out of business. They were forced plumb out of business, because they didn't have any land. They didn't have any base. You had to have a base.

MB: How did the Taylor Grazing Act affect your family?

SC: Oh, we never had any problems. We had a big base. My dad had accumulated a big land base. It was strung out over a long, long area, so we never suffered a bit from the Taylor Grazing. Fact is, they enhanced us, because it got rid of those coyote outfits, and in later years, we got a adjudication, and got our own allotments, to where we could manage our own grass. When it first come in, you know, it was open area. You still had a permit out there, but it was community...grazing, you know. Then it was adjudicated into permits or allotments. Then you could control your own feed.

MB: How...when you tell me about all this, and all the knowledge you gained from how to manage the sheep—where did you learn most of that?

SC: Just by experience. Of course, I had a minor in range management from the university, and a lot of botany, and I knew the plants, the palatability of them, and had some grazing, but that was a very...in those days, the grazing was a very inaccurate science, and it's still not very accurate, but it's come a long ways in the last forty years, the science of grazing. My dad was just a natural land manager, as far as grazing was concerned.

MB: What do you mean by that?

SC: He just knew how to take care of that, to where he'd get the maximum production out of it, and he always had more grass than he could use. So if some disaster come along, like a fire or snowstorm, or something, he always had...he had enough feed. You know, you're always on those ranges, especially your foothill ranges, you're always faced with the prospect of fire, any year.

And so, he never had enough of it... we always had enough range that we could... we didn't need to use it all any given year.

MB: Hmmm, that's interesting. Did you learn a lot, then, from your father on how the range...?

SC: Oh yeah, and by just experience yourself, you know. Over the years you pick up a lot of that.

BIBLIOGRAPHY

Bibliography of sources not otherwise identified in the text or for which complete bibliographic information was not supplied.

Davis, H. L. *Honey in the Horn*. New York: Harper and Brothers, 1935.

———. *Winds of Morning*. New York: William Morrow, 1952.

———. *Team Bells Woke Me, and Other Stories*. New York: William Morrow, 1953.

Doig, Ivan. *This House of Sky: Landscape of the Western Mind*. New York: Harcourt, Brace, Jovanovich, 1978.

———. *English Creek*. New York: Atheneum, 1984.

———. *Dancing at the Rascal Fair*. New York: Atheneum, 1987.

Douglas, William A. *Basque Sheepherders of the American West*. Reno: University of Nevada Press, 1985.

Ehrlich, Gretel. *The Solace of Open Spaces*. New York: Viking, 1985.

Gemming, Elizabeth. *Wool Gathering: Sheep Raising in Old New England*. New York: Coward, McCann and Geohagen, 1979.

Gregg, Jacob Ray. *Pioneer Days in Malheur County*. Los Angeles: Lorin L. Morrison, 1950.

Hanley, Mike, with Omer Stanford. *Sage Brush and Axle Grease*. Jordan Valley, Ore.: Mike Hanley, 1976.

Jordan, Grace. *Home Below Hell's Canyon*. New York: Crowell, 1954.

Oberg, Pearl M. *Between These Mountains*. New York: Exposition Press, 1970.

Sherlock, Patti. *Alone on the Mountain*. Garden City, New York: Doubleday & Company, Inc., 1979.

Steiner, Stan. *The Ranchers: A Book of Generations*. New York: Alfred A. Knopf, 1980.

———. *The Waning of the West*. St. Martin's Press, 1989.

Stockton, Bill. *Today I Baled Some Hay to Feed the Sheep the Coyotes Eat*. Helena: Falcon Press Publishing Co. Inc., 1983.

Yensen, Dana. *A Grazing History of Southwestern Idaho with Emphasis on the Birds of Prey Study Area*. Moscow, Idaho: Department of Biological Sciences, University of Idaho, 1980.

Young, James A., with B. Abbott Sparks. *Cattle in the Cold Desert*. Logan, Utah: Utah State University Press, 1985.

Index

Aguirre, Frank, 29, 30, 37, 67, 72–74
Aguirre, Mary Ann, 37
Aguirre, Mrs. Tony, 37
Albertson College of Idaho, 6, 12
And the Ladies of the Club (Santmeyer), 15
Anderson, Dean, 93
Anderson, Scott, 114
Armstrong, Charles, 69
As You Like It (Shakespeare), 1–2
Attebery, Jennifer Eastman, 115
Attebery, Louie W., 29, 94
Australian shearers, 36

"Back at the Ranch" (Robbins), 2
Baker, Dr. E. T., 7, 25
Baker, Oreg., 76, 77
Ballad of Josie, The (movie), 27
Bartlett, Dr., 76, 78
Basque herders, 29, 79–80, 94, 103, 120, 122
Beeson, Judd (John), 72, 73, 74
Belled animals, 9, 12, 72, 74
Boyd, Stan, 93
Boyd, Tom, 73, 74, 76
Braly, David, 98
Bruce, Jack, 113
Brunvand, Jan, 101
Buckendorf, Madeline, 80, 81, 117–25
Bucks. *See* Sheep industry
Bureau of Land Management, 13–14, 16, 100

Caldwell, Idaho, 11
Campbell, Margaret, 37
Camps, 62–67, 72, 73–74, 123
Camptenders: at work, 9, 27, 29, 59, 74; duties, 62, 67; socializing, 38; tales of, 74–75
Casmyer, Victor, 28
Cattle industry, 7, 27, 93–97, 98, 99
"Celts and Other Folk in the Regional Livestock Industry," 27

Chambers, Gale, 95
Clyde, Erlene (Mrs. Phil Soulen), 6–7
Coles, Mrs. Jeanne (Poulton), 105
Coles, W. R., 105
Colorado Springs, Colo., 15
"Comeback in Cattle Country, The" (Braly), 98
Computers in ranching, 2
Corta, Ella, 28
Council, Idaho, 95
Counting of sheep, 37–38, 68, 72
Counting Sheep (McGregor), 33–34. 62
Cowboys, 27, 96–97. *See also* Cattle industry
Cruickshank, Alex, 113
Cruickshank, Stewart, 80–81, 117–25
Cummings, Daniel, 95

Davis, Diamondfield Jack, 95
Day, Doris, 28
"Dead Line," 95
Diseases: "bumblefoot" in sheep, 12, 26; during lambing, 25; foot rot, 10, 12, 26–27, 30; in range sheep, 69; lameness in sheep, 12; lumpjaw in sheep, 16; poisoning by parsnip roots, 112; protection from, 106, 108; Rocky Mountain spotted fever in herders, 28; "stiff lambs," 10; worms in sheep, 107. *See also* Sheep industry
Drewsey, Oreg., 33, 95
Dry Lake, Idaho, 11
Dundes, Alan, 3
Dynamics of Folklore (Toelken), 3

Embryo technology, 2, 5, 98–99
Emmett, Idaho, 117–18
Ewes. *See* Sheep industry
Examiner, Emmett, Idaho, 111, 113

Farming and Democracy (Griswold), 5
Farnsworth, Richard, 2

"Fiddle Tune, The" (Attebery), 101
Fleeces, 31, 36 , 73, 108. *See also* Wool industry
Fleming family, 79n
Folklore, 3, 28–29, 70–72, 74–78, 94–95
"Folklore of the Lower Snake River Valley" (Attebery), 29, 94
Fraser family, 79n
French herders, 28
From Here to Eternity (Jones), 15
Frost, Robert, 4

German herders, 122
Gold, 29
Graham, Bob, 63, 66
Graham, Idaho, 63
Grasses on rangeland, 16
Graves, Peter, 27
Grazing, 13, 16, 109–10. *See also* Sheep industry; Taylor Grazing Act
Greek herders, 29
Griswold, A. Whitney, 4–5

Hailey, John, 79n
Hart, Arthur, 79n
Hartley, Ellis, 95
Herders: camptenders and, 67; duties, 67–70; in legends, 27–29; life and work, 9, 12, 13, 59–81, 120–21; nationalities of, 13, 27–29, 66, 72–80, 94, 103, 120, 122; social life, 38, 70; wages, 79–80, 123. *See also* Sheepmen
Hillinger, Charles, 79, 80
Hogs as camp provisions, 30, 62–63
Holmes, Fred, 79
Horses and range damage, 96
Huntington Land and Livestock, 76–77

Idaho Farmer-Stockman, 7
Idaho Press Tribune, 98
Idaho Statesman, 79, 93, 103
Idaho Wool Growers Association, 93
Idaho Yesterdays, 27
Idaho-Stockmen Farmer, 95
Index, Emmett, Idaho, 111–14
Irish herders, 27, 122
Italian herders, 122

Jayo, Eusebio, 80
Jones, Dr., 28
Jones, James, 15
Jones, Suzi, 97
Jordan Valley, Oreg., 28, 62
Joyce, Martin, 95

Kasmeyer, Victor, 28
Kennedy, Hugh, 94

Lackey, John, 94
Laidlaw family, 79n
Lake Lowell, Idaho, 12
Lambing: in New Zealand, 105–6; in sheds, 25, 119; on range, 8, 10, 38–40, 119; timing, 4, 12, 111. *See also* Sheep industry
Lambs. *See* Sheep industry
Letha, Idaho, 12, 25, 27, 30–31, 37–38, 61, 74
Life magazine, 79
Lightening strikes, 69
Little family, 79n, 111, 112, 113, 114, 118
Lopez, Joe, 34
Lost Sheepherder Mine, 28, 76
Lovell, Gary, 36

McBride, Oreg., 94
McCall, Idaho, 115
McConnel brothers, 113
McGinnis, Glen, 28
McGregor, Alexander Campbell, 34–35, 44, 62, 68, 101
McMillan, John, 79n
McMullen, Orren, 33
Melba, Idaho, 11
Mexican shearers, 34, 112, 122
Midvale Market, Midvale, Idaho, 63, 66
Military establishments, 15–16
Miller, Dick, 97
Moscow, Idaho, 6

New Zealand sheep industry, 36, 105–10, 122
Northwest Folklore, 72
Norwegian herders, 122
Nyssa, Oreg., 28

Olafson, Magni, 27, 60
Oregonian, The, 98

Palmer, John, 94

Parma, Idaho, 117
Peruvian herders, 29, 66, 72–74, 79, 122
Pickle Butte, Idaho, 11, 12
Portuguese herders, 122
Predators, 111; bears, 10, 43, 69, 70, 75; bobcats, 10, 69; birds, 69; coyotes, 10, 14, 39, 43, 68–69, 70, 111; dogs, 69–70
Proctor, David, 103
Public lands, 99–100

Rambaud, Frank, 28
Rambaud, Pierre "Pete," 28
"Range Sheep Management" (Soulen), 7–10
Rangeland, 16, 96, 99–100
Readings in American Folklore (Brunvand), 101
Robbins, Steve, 2
Rocky Mountain spotted fever, 28
Russian herders, 122

Salt for sheep bands, 74
Sanderson, Stewart, 59
Santmeyer, Helen Hoover, 15
Scots herders, 27, 66, 79, 120, 122
Scott, George, 79n
Shakespeare, William, 1
Shea, Con, 94
Shearing: in New Zealand, 108–9; modern methods, 31–33; nationalities of shearers, 34, 36, 112, 122; sheep-shearing implements, 35–36; timing of, 4, 9, 12, 25, 112; traditional methods, 33–37; women shearers, 36. *See also* Sheep industry
Sheep and cattle wars, 93–97
Sheep breeds: Columbia, 26, 103; Corriedale, 8, 25; Hampshire, 9, 40; Panama, 8, 25, 26, 98, 103; Polypay, 103; Rambouillet, 25–26; Romney, 8, 25; Suffolk, 9, 26, 98; Targhee, 103
Sheep camps. *See* Camps
Sheep dogs, 9, 12, 13, 59, 67, 68, 121
Sheep industry: artificial insemination, 2; bands of sheep, 8, 9, 12, 61–62, 120; bucks, 8, 9, 25, 30–31; castrating sheep, 42, 44–45, 79, 106; counting sheep, 37–38, 68, 72; "coyote outfits," 13–14, 16, 123; dagging sheep, 109; embryo technology, 2, 5, 98–99; feeding sheep, 3–4, 8–10, 11, 74; grafting lambs, 40, 43–44; grazing, 13, 16, 109–10; in New Zealand, 36, 105–10, 122; lamb consumption, 93, 104, 107–8; marking sheep, 38, 40–45; overview, 2, 103–4; press coverage, 79; range management, 7–17, 96, 99–100; Stewart Cruickshank's story, 117–25; tail docking of sheep, 42–43, 106; twinning, 39. *See also* Sheep breeds; Diseases; Lambing; Predators; Shearing; Wool industry
Sheep Shearers union, 34
Sheepherders. *See* Herders
Sheepmen, 13–14, 16, 93–97, 123. *See also* Herders
Skillerus family, 79n
Skinner, Bob, 62
Smith, Harry, 99
Smith, John, 95
Soulen family enterprise, 101
Soulen family home, 115
Soulen Livestock Company, 6, 7, 62, 63
Soulen, Beulah May (Mrs. Harry B.), xv, xvi, xviii, 6, 115
Soulen, Harry Boone: life history, xiii–xix; on herders, 59; on "hot vitriol" treatment, 26; on lambing, 25; on raising sheep, 6, 7–10, 11, 13
Soulen, Harry (Harry B.'s grandson), 7, 37, 44, 72–74
Soulen, Phil (Harry B.'s son): as cattleman and sheepman, 97; his camps, 11–13, 72–74, 76; his grazing areas, 16–17; his training, 6–7, 44, 61; on castration, 79; on dual-capability sheep, 25–26; on lambing, 25, 38–41; on shearing, 30, 31; on wool "trompers," 36; treating foot diseases, 26; working sheep, 37
Spanish herders, 29
Spokane, Wash., 33
Stanfield, Senator, 74, 75, 76–78
Stanford, Dr. Lyle, 14
Storytelling. *See* Folklore
Stringer, John, 74, 75, 76–78
Study of Folklore, The (Dundes), 3
Sugar Mountain (camptender), 74, 76
Swedish herders, 27

Taylor Grazing Act, 13, 28, 99, 103, 123–24. *See also* Grazing
Toelken, Barre, 3
Transhumance, 3, 4, 5, 29, 30, 101

U.S. Department of Agriculture, 93
U.S. Department of Immigration, 13
U.S. Department of Interior, 13
U.S. Department of Labor, 13
U.S. Forest Service, 72
University of Idaho, 98, 118
University of Nebraska, 2

Vale, Oreg., 28
Valle, Vincente, 79–80
Van Duesen family, 113, 118
Vanishing Kingdom (McMullen), 33

Water supplies, 11
Webfoots and Bunchgrassers (Jones), 97
Weiser, Idaho, 6, 7, 40
Welch, James, 79n
Western Range Association, 13
"Whose Home on the Range?" (Congress), 5
Wilson, Al, 27
Wilson, John, 95
Wing, Jack, 79
Wool industry: contamination of wool, 42; early history, 79n, 114; fleeces, 31, 36, 73, 108; income from wool, 104; market for wool, 93; sheep breeds for wool, 25–26; woolen cloth, 74

Yeutter, Clayton, 2